Steven Weinberg

DREAMS OF A FINAL THEORY

Steven Weinberg received the 1979 Nobel Prize for Physics for his work in unifying two of the fundamental forces of nature, and in 1991 he was awarded the National Medal of Science at the White House. His earlier prize-winning book *The First Three Minutes* is the classic account of the "Big Bang," the modern theory of the origin of the universe. Among his other books are *The Theory of Subatomic Particles* and *Gravitation and Cosmology: Principles and Applications of the General Theory of Relativity*. Steven Weinberg is a member of the Royal Society of London as well as the U.S. National Academy of Sciences, and has been awarded numerous honorary degrees, most recently from Columbia University and the Universities of Salamanca and Padua.

Also by
Steven Weinberg

The First Three Minutes
The Discovery of Subatomic Particles
Elementary Particles and the Laws of Physics (with Richard Feynman)
Gravitation and Cosmology

3932
SE WOODSTOCK

"I studied no science of any kind after tenth-grade chemistry. One of the virtues of Weinberg's book is that it is able to communicate to a scientific ignoramus, and does so because of the [qualities of] detachment, patience, and clarity, organically connected by deep commitment to the subject. As it happens, for Weinberg ... beauty of an objective sort is central to his commitment."   —Mindy Aloof, *Atlantic Monthly*

"*Dreams of a Final Theory* is a good book and an honest one."
—Phillip Johnson, *Wall Street Journal*

"He moves beyond the simplistic description of experimenters doing their job to prove or disprove theories to give us a much richer picture of how science actually works."
—Jon Van, *Chicago Tribune*

"The physics world is bound to have high expectations for a new book by Steven Weinberg, one of its most distinguished citizens. *Dreams of a Final Theory* does not disappoint. ... I came away from a first reading of the book eager to read it again. A second reading left me mightily impressed with its subtlety and honesty. ... *Dreams of a Final Theory* deserves to be read and re-read by thoughtful physicists, philosophers, and just plain thoughtful people."   —Frank Wilczek, *Physics Today*

"A compelling plea for seeking the 'final laws of nature.'"
—Sharon Begley, *Newsweek*

"He writes clearly and with confidence, imbuing the reader with an irresistible sense that one is in the hands of a master physicist at play."
—Michael White, *Sunday Times* (London)

DREAMS OF A
FINAL THEORY

Steven Weinberg

Vintage Books
A Division of Random House, Inc.
New York

To Louise and Elizabeth

FIRST VINTAGE BOOKS EDITION, JANUARY 1994

Library of Congress Cataloging-in-Publication Data
Weinberg, Steven, 1933–
Dreams of a final theory / Steven Weinberg.
p. cm.
Originally published: New York: Pantheon Books, 1992.
Includes bibliographical references and index.
ISBN 0–679–74408–8
1. Physics. 2. Relativity (Physics) 3. Quantum theory.
I. Title.
QC21.2.W428 194
530—dc20   93-30534
CIP

Book design by Fearn Cutler

Manufactured in the United States of America

10   9   8   7   6   5

CONTENTS

PREFACE

This book is about a great intellectual adventure, the search for the final laws of nature. The dream of a final theory inspires much of today's work in high-energy physics, and though we do not know what the final laws might be or how many years will pass before they are discovered, already in today's theories we think we are beginning to catch glimpses of the outlines of a final theory.

The very idea of a final theory is controversial, and the subject of intense current debate. This controversy has even reached the committee rooms of Congress; high-energy physics has become increasingly expensive, and its claims to public support hinge in part on its historical mission of uncovering the final laws.

From the beginning my intention here has been to lay out the issues raised by the idea of a final theory as part of the intellectual history of our times, for readers with no prior knowledge of physics or higher mathematics. This book does touch on the key ideas that underlie today's work at the frontier of physics. But this is not a physics textbook, and the reader will not encounter neat separate chapters here on particles, forces, symmetries, and strings. Instead I have woven the concepts of modern physics into the discussion of what we mean by a final theory and how we are going to find it. In this I have been guided by my own experience as a reader in fields, such as history, in which *I* am an outsider. Historians often succumb to the temptation of first giving a narrative history, followed by separate background chapters on population, economics, technology, and so on. On the other hand, the historians one reads for pleasure, from Tacitus and Gibbon to J. H. Elliott and S. E. Morison, mingle narrative and background while at the same time making a case for any conclusions that they wish to set before the reader. In writing this book I have tried to follow their lead, and to resist the temptations of tidiness. I have also not hesitated to bring in historical or scientific material that may be already familiar to readers who are historians or scientists, or even to repeat this material where I thought it would be useful. As Enrico Fermi once said, one should never underestimate the pleasure we feel from hearing something we already know.

Dreams of a Final Theory is roughly divided into three parts and a coda. The first part, chapters 1 through 3, presents the idea of a final theory; chapters 4 through 8 explain how we have been able to make progress toward a final theory; and chapters 9 through 11 look ahead to speculate about the shape of the final theory, and about how its discovery will affect humankind. Finally, in chapter 12 I turn to the arguments for and against the Superconducting Super Collider, an expensive new instrument that is desperately needed by high-energy physicists but whose future funding remains in doubt.

Readers will find a fuller discussion of some of the ideas in the main text in a series of notes at the back of the book. In some places, where I have had to oversimplify some scientific concept in the main text, I have given a more precise discussion in an endnote. These endnotes also include bibliographical references for some of the material quoted in the text.

I am deeply grateful to Louise Weinberg for urging me to rewrite an earlier version of this book, and for seeing how it should be done.

My warm thanks go to Dan Frank of Pantheon Books for his encouragement and his perceptive guidance and editing, and to Neil Belton of Hutchinson Radius and my agent, Morton Janklow, for important suggestions.

I am also indebted for comments and advice on various topics to the philosophers Paul Feyerabend, George Gale, Sandra Harding, Myles Jackson, Robert Nozick, Hilary Putnam, and Michael Redhead; the historians Stephen Brush, Peter Green, and Robert Hankinson; the legal scholars Philip Bobbitt, Louise Weinberg, and Mark Yudof; the physicist-historians Gerald Holton, Abraham Pais, and S. Samuel Schweber; the physicist-theologian John Polkinghorne; the psychiatrists Leon Eisenberg and Elizabeth Weinberg; the biologists Sydney Brenner, Francis Crick, Lawrence Gilbert, Stephen J. Gould, and Ernst Mayr; the physicists Yakir Aharonov, Sidney Coleman, Bryce De Witt, Manfred Fink, Michael Fisher, David Gross, Bengt Nagel, Stephen Orzsag, Brian Pippard, Joseph Polchinski, Roy Schwitters, and Leonard Susskind; the chemist Roald Hoffmann; the astrophysicists William Press, Paul Shapiro, and Ethan Vishniac; and the writers James Gleick and Lars Gustafsson. Many serious errors were avoided with their help.

Steven Weinberg
Austin, Texas
August 1992

DREAMS OF A
FINAL THEORY

CHAPTER I

PROLOGUE

If ever any beauty I did see,
Which I desir'd, and got, 'twas but a dream of thee.

John Donne, *The Good-Morrow*

T he century now coming to a close has seen in physics a dazzling expansion of the frontiers of scientific knowledge. Einstein's special and general theories of relativity have permanently changed our view of space and time and gravitation. In an even more radical break with the past, quantum mechanics has transformed the very language we use to describe nature: in place of particles with definite positions and velocities, we have learned to speak of wave functions and probabilities. Out of the fusion of relativity with quantum mechanics there has evolved a new view of the world, one in which matter has lost its central role. This role has been usurped by principles of symmetry, some of them hidden from view in the present state of the universe. On this foundation we have built a successful theory of electromagnetism and the weak and strong nuclear interactions of elementary particles. Often we have felt as did Siegfried after he tasted the dragon's blood, when he found to his surprise that he could understand the language of birds.

But now we are stuck. The years since the mid-1970s have been the most frustrating in the history of elementary particle physics. We are paying the price of our own success: theory has advanced so far that further progress will require the study of processes at energies beyond the reach of existing experimental facilities.

In order to break out of this impasse, physicists began in 1982 to develop plans for a scientific project of unprecedented size and cost, known as the Superconducting Super Collider. The plan in its final form called for a 53-mile-long oval tunnel to be dug at a site south of Dallas. Within this underground tunnel thousands of superconducting magnets would guide two beams of electrically charged particles known as protons in opposite directions millions of times around the ring, while the protons would be accelerated to an energy twenty times larger than the highest energy achieved at existing particle accelerators. At several points along the ring the protons in the two beams would be made to collide hundreds of millions of times a second, and enormous detectors, some weighing tens of thousands of tons, would record what happens in these collisions. The cost of the project was estimated at over 8 billion dollars.

The Super Collider has attracted intense opposition, not only from frugal congressmen but also from some scientists who would rather see the money spent in their own fields. There is much grumbling about so-called big science, and some of it has found a target in the Super Collider. Meanwhile, the European consortium known as CERN is considering the construction of a somewhat similar facility, the Large Hadron Collider, or LHC. The LHC would cost less than the Super Collider, because it would make use of an existing tunnel under the Jura Mountains near Geneva, but for the same reason its energy would be limited to less than half that of the Super Collider. In many respects the American debate over the Super Collider is paralleled by a debate in Europe over whether to build the LHC.

As this book goes to press in 1992, funding for the Super Collider, which was cut off by a June vote in the House of Representatives, has been restored by an August vote in the Senate. The future of the Super Collider would be assured if it received appreciable foreign support, but so far that has not been forthcoming. As matters stand, even though funding for the Super Collider has survived in Congress this year, it faces the possibility of cancellation by Congress next year, and in each year until the project is completed. It may be that the closing years of the twentieth century will see the epochal search for the foundations of physical science come to a stop, perhaps only to be resumed many years later.

This is not a book about the Super Collider. But the debate over the project has forced me in public talks and in testimony before Congress to try to explain what we are trying to accomplish in our studies of elementary particles. One might think that after thirty years of work as a physicist I would have no trouble with this, but it turns out to be not so easy.

For myself, the pleasure of the work had always provided justification enough for doing it. Sitting at my desk or at some café table, I manipulate mathematical expressions and feel like Faust playing with his pentagrams before Mephistopheles arrives. Every once in a while mathematical abstractions, experimental data, and physical intuition come together in a definite theory about particles and forces and symmetries. And every once in an even longer while the theory turns out to be right; sometimes experiments show that nature really does behave the way the theory says it ought.

But this is not all. For physicists whose work deals with elementary particles, there is another motivation, one that is harder to explain, even to ourselves.

Our present theories are of only limited validity, still tentative and incomplete. But behind them now and then we catch glimpses of a final theory, one that would be of unlimited validity and entirely satisfying in its completeness and consistency. We search for universal truths about nature, and, when we find them, we attempt to explain them by showing how they can be deduced from deeper truths. Think of the space of scientific principles as being filled with arrows, pointing toward each principle and away from the others by which it is explained. These arrows of explanation have already revealed a remarkable pattern: they do not form separate disconnected clumps, representing independent sciences, and they do not wander aimlessly—rather they are all connected, and if followed backward they all seem to flow from a common starting point. This starting point, to which all explanations may be traced, is what I mean by a final theory.

We certainly do not have a final theory yet, and we are not likely to discover it soon. But from time to time we catch hints that it is not so very far off. Sometimes in discussions among physicists, when it turns out that mathematically beautiful ideas are actually relevant to the real world, we get the feeling that there is something behind the blackboard, some deeper truth foreshadowing a final theory that makes our ideas turn out so well.

Speaking of a final theory, a thousand questions and qualifications crowd into the mind. What do we mean by one scientific principle "explaining" another? How do we know that there is a common starting point for all such explanations? Will we ever discover that point? How close are we now? What will the final theory be like? What parts of our present physics will survive in a final theory? What will it say about life and consciousness? And, when we have our final theory, what will happen to science and to the human spirit? This chapter, barely touching on these questions, leaves a fuller response to the rest of this book.

The dream of a final theory did not start in the twentieth century. It may be traced in the West back to a school that flourished a century before the birth of Socrates in the Greek town of Miletus, where the Meander River empties into the Aegean Sea. We do not really know much about what the pre-Socratics taught, but later accounts and the few original fragments that survive suggest that the Milesians were already searching for explanations of all natural phenomena in terms of a fundamental constituent of matter. For Thales, the first of these Milesians, the fundamental substance was water; for Anaximenes, the last of this school, it was air.

Today Thales and Anaximenes seem quaint. Much more admiration is given now to a school that grew up a century later at Abdera on the seacoast of Thrace. There Democritus and Leucippus taught that all matter is composed of tiny eternal particles they called atoms. (Atomism has roots in Indian metaphysics that go back even earlier than Democritus and Leucippus.) These early atomists may seem wonderfully precocious, but it does not seem to me very important that the Milesians were "wrong" and that the atomic theory of Democritus and Leucippus was in some sense "right." None of the pre-Socratics, neither at Miletus nor at Abdera, had anything like our modern idea of what a successful scientific explanation would have to accomplish: the *quantitative* understanding of phenomena. How far do we progress toward understanding why nature is the way it is if Thales or Democritus tells us that a stone is made of water or atoms, when we still do not know how to calculate its density or hardness or electrical conductivity? And of course, without the capacity for quantitative prediction, we could never tell whether Thales or Democritus is right.

On the occasions when at Texas and Harvard I have taught physics to liberal-arts undergraduates, I have felt that my most important task (and certainly the most difficult) was to give the students a taste of the power of being able to calculate in detail what happens under various circumstances in various physical systems. They were taught to calculate the deflection of a cathode ray or the fall of an oil droplet, not because that is the sort of thing everyone needs to calculate but because in doing these calculations they could experience for themselves what the principles of physics really mean. Our knowledge of the principles that determine these and other motions is at the core of physical science and a precious part of our civilization.

From this point of view, the "physics" of Aristotle was no better than the earlier and less sophisticated speculations of Thales and Democritus. In his books *Physics* and *On the Heavens* Aristotle describes the motion of a projectile as being partly natural and partly unnatural; its natural motion, as for all heavy bodies, is downward, toward the center of things, and its unnatural motion is imparted by the air, whose motion can be traced to whatever started the projectile in motion. But just how fast does the projectile travel along its path, and how far does it get before it hits the ground? Aristotle does not say that the calculation or measurements are too difficult or that not enough is yet known about the laws of motion to give a detailed description of the projectile's motion. Rather, Aristotle does not offer an answer, right or wrong, because he does not realize that these are questions worth asking.

And why are they worth asking? The reader, like Aristotle, might not care very much how fast the projectile falls—I don't much care myself. The important thing is that we now know the *principles*—Newton's law of motion and gravitation and the equations of aerodynamics—that determine precisely where the projectile is at every moment in its flight. I am not saying here that we actually can calculate exactly how the projectile moves. The flow of air past an irregular stone or an arrow's feathers is complicated, and so our calculations are likely to be only fair approximations, especially for air flows that become turbulent. There is also the problem of specifying precise initial conditions. Nevertheless we can use our known physical principles to solve simpler problems, like the motion of planets in airless space or the steady flow of air around spheres or plates, well enough to reassure us that we really do know what principles govern the projectile's flight. In the same way, we cannot calculate the course of biological evolution, but we now know pretty well the principles by which it is governed.

This is an important distinction, one that tends to get muddled in arguments over the meaning or the existence of final laws of nature. When we say that one truth explains another, as for instance that the physical principles (the rules of quantum mechanics) governing electrons in electric fields explain the laws of chemistry, we do not necessarily mean that we can actually deduce the truths we claim have been explained. Sometimes we can complete the deduction, as for the chemistry of the very simple hydrogen molecule. But sometimes the problem is just too complicated for us. In speaking in this way of scientific explanations, we have in mind not what scientists actually deduce but instead a necessity built into nature itself. For instance, even before physicists and astronomers learned in the nineteenth century how to take account of the mutual attraction of the planets in accurate calculations of their motions, they could be reasonably sure that the planets move the way they do because they are governed by Newton's laws of motion and gravitation, or whatever more exact laws Newton's laws approximate. Today, even though we cannot predict everything that chemists may observe, we believe that atoms behave the way they do in chemical reactions because the physical principles that govern the electrons and electric forces inside atoms leave no freedom for the atoms to behave in any other way.

This is a tricky point in part because it is awkward to talk about one fact explaining another without real people actually doing the deductions. But I think that we have to talk this way because this is what our science is *about:* the discovery of explanations built into the logical structure of nature. Of course we become much more confident that we have the correct explanation when we are able actually to carry out *some* calculations and compare the results with observation: if not of the chemistry of proteins, then at least of the chemistry of hydrogen.

Even though the Greeks did not have our goal of a comprehensive and quantitative understanding of nature, precise quantitative reasoning was certainly not unknown in the ancient world. For millennia people have known about the rules of arithmetic and plane geometry and the grand periodicities of the sun and moon and stars, including even such subtleties as the precision of the equinoxes. Beyond this, there was a flowering of mathematical science after Aristotle, during the Hellenistic era that spans the time from the conquests of Aristotle's pupil Alexander to the subjugation of the Greek world by Rome. As an undergraduate studying philosophy I felt some pain at hearing Hellenic philosophers like Thales or Democritus called physicists; but, when we came to the great Hellenistics, to Archimedes in Syracuse discovering the laws of buoyancy or Eratosthenes in Alexandria measuring the circumference of the earth, I felt at home among my fellow scientists. Nothing like Hellenistic science was seen anywhere in the world until the rise of modern science in Europe in the seventeenth century.

Yet for all their brilliance, the Hellenistic natural philosophers never came close to the idea of a body of laws that would precisely regulate *all* nature. Indeed, the word "law" was rarely used in antiquity (and never by Aristotle) except in its original sense, of human or divine laws governing human conduct. (It is true that the word "astronomy" derives from the Greek words *astron* for a star and *nomos* for law, but this term was less often used in antiquity for the science of the heavens than the word "astrology.") Not until Galileo, Kepler, and Descartes in the seventeenth century do we find the modern notion of laws of nature.

The classicist Peter Green blames the limitations of Greek science in large part on the persistent intellectual snobbery of the Greeks, with their preference for the static over the dynamic and for contemplation over technology, except for military technology. The first three kings of Hellenistic Alexandria supported research on the flight of projectiles because of its military applications, but to the Greeks it would have seemed inappropriate to apply precise reasoning to something as banal as the process by which a ball rolls down an inclined plane, the problem that illuminated the laws of motion for Galileo. Modern science has its own snobberies—biologists pay more attention to genes than to bunions, and physicists would rather study proton-proton collisions at 20 trillion volts than at 20 volts. But these are tactical snobberies, based on judgments (right or wrong) that some phenomena turn out to be more revealing than others; they do not reflect a conviction that some phenomena are more important than others.

It is with Isaac Newton that the modern dream of a final theory really begins. Quantitative scientific reasoning had never really disappeared, and by Newton's time it had already been revitalized, most notably by Galileo. But Newton was able to explain so much with his laws of motion and law of universal gravitation, from the orbits of planets and moons to the rise and fall of tides and apples, that he must for the first time have sensed the possibility of a really comprehensive explanatory theory. Newton's hopes were expressed in the preface to the first edition of his great book, the *Principia:* "I wish we could derive the rest of the phenomena of nature [that is, the phenomena not treated in the *Principia*] by the same kind of reasoning as for mechanical principles. For I am induced by many reasons to suspect that they may all depend on certain forces." Twenty years later, Newton described in the *Opticks* how he thought his program might be carried out:

> Now the smallest particles of matter cohere by the strongest attractions, and compose bigger particles of weaker virtue; and many of these may cohere and compose bigger particles whose virtue is still weaker, and so on for diverse successions, until the progression ends in the biggest particles on which the operations in chemistry, and the colours of natural bodies depend, and which by cohering compose bodies of a sensible magnitude. There are therefore agents in nature able to make the particles of bodies stick together by very strong attractions. And it is the business of experimental philosophy to find them out.

Newton's great example gave rise especially in England to a characteristic style of scientific explanation: matter is conceived to consist of tiny immutable particles; the particles act on one another through "certain forces," of which gravitation is just one variety; knowing the positions and velocities of these particles at any one instant, and knowing how to calculate the forces among them, one can use the laws of motion to predict where they will be at any later time. Physics is often still taught to freshmen in this fashion. Regrettably, despite the further successes of physics in this Newtonian style, it was a dead end.

After all, the world is a complicated place. As scientists learned more about chemistry and light and electricity and heat in the eighteenth and nineteenth centuries, the possibility of an explanation along Newtonian lines must have seemed more and more remote. In particular, in order to explain chemical reactions and affinities by treating atoms as Newtonian particles moving under the influence of their mutual attraction and repulsion, physicists would have had to make so many arbitrary assumptions about atoms and forces that nothing really could have been accomplished.

Nevertheless, by the 1890s an odd sense of completion had spread to many scientists. In the folklore of science there is an apocryphal story about some physicist who, near the turn of the century, proclaimed that physics was just about complete, with nothing left to be done but to carry measurements to a few more decimal places. The story seems to originate in a remark made in 1894 in a talk at the University of Chicago by the American experimental physicist Albert Michelson: "While it is never safe to affirm that the future of Physical Science has no marvels in store even more astonishing than those of the past, it seems probable that most of the grand underlying principles have been firmly established and that further advances are to be sought chiefly in the rigorous application of these principles to all the phenomena which come under our notice. . . . An eminent physicist has remarked that the future truths of Physical Science are to be looked for in the sixth place of decimals." Robert Andrews Millikan, another American experimentalist, was in the audience at Chicago during Michelson's talk and guessed that the "eminent physicist" Michelson referred to was the influential Scot, William Thomson, Lord Kelvin. A friend has told me that when he was a student at Cambridge in the late 1940s, Kelvin was widely quoted as having said that there was nothing new to be discovered in physics and that all that remained was more and more precise measurement.

I have not been able to find this remark in Kelvin's collected speeches, but there is plenty of other evidence for a widespread, though not universal, sense of scientific complacency in the late nineteenth century. When the young Max Planck entered the

University of Munich in 1875, the professor of physics, Philip Jolly, urged him against studying science. In Jolly's view there was nothing left to be discovered. Millikan received similar advice: "In 1894," he recalled, "I lived in a fifth-floor flat on Sixty-fourth Street, a block west of Broadway, with four other Columbia graduate students, one a medic and the other three working in sociology and political science, and I was ragged continuously by all of them for sticking to a 'finished,' yes, a 'dead subject,' like physics, when the new, 'live' field of the social sciences was just opening up."

Often such examples of nineteenth-century complacency are trotted out as warnings to those of us in the twentieth century who dare to talk of a final theory. This rather misses the point of these self-satisfied remarks. Michelson and Jolly and Millikan's roommates could not possibly have thought that the nature of chemical attraction had been successfully explained by physicists—much less that the mechanism of heredity had been successfully explained by chemists. Those who made such remarks could only have done so because they had given up on the old dream of Newton and his followers that chemistry and all other sciences would be understood in terms of physical forces; for them, chemistry and physics had become co-equal sciences, each separately near completion. To whatever extent there was a widespread sense of completeness in late-nineteenth-century science, it represented only the complacency that comes with diminished ambition.

But things were to change very rapidly. To a physicist the twentieth century begins in 1895, with Wilhelm Roentgen's unexpected discovery of X rays. It was not that X rays themselves were so important; rather, their discovery encouraged physicists to believe that there were many new things to be discovered, especially by studying radiation of various sorts. And discoveries did follow in quick succession. At Paris in 1896 Henri Becquerel discovered radioactivity. At Cambridge in 1897 J. J.

Thomson measured the deflection of cathode rays by electric and magnetic fields and interpreted the results in terms of a fundamental particle, the electron, present in all matter, not only in cathode rays. At Bern in 1905 Albert Einstein (while still excluded from academic employment) presented a new view of space and time in his special theory of relativity, suggested a new way of demonstrating the existence of atoms, and interpreted earlier work of Max Planck on heat radiation in terms of a new elementary particle, the particle of light later called the photon. A little later, in 1911, Ernest Rutherford used the results of experiments with radioactive elements in his Manchester laboratory to infer that atoms consist of small massive nuclei surrounded by clouds of electrons. And in 1913 the Dane Niels Bohr used this atomic model and Einstein's photon idea to explain the spectrum of the simplest atom, that of hydrogen. Complacency gave way to excitement; physicists began to feel that a final theory unifying at least all physical science might soon be found.

Already in 1902, the previously complacent Michelson could proclaim: "The day appears not far distant when the converging lines from many apparently remote regions of thought will meet on ... common ground. Then the nature of the atoms, and the forces called into play in their chemical union; the interactions between these atoms and the non-differentiated ether as manifested in the phenomenon of light and electricity; the structures of the molecules and molecular systems of which the atoms are the units; the explanation of cohesion, elasticity, and gravitation—all these will be marshalled into a single and compact body of scientific knowledge." Where before Michelson had thought that physics was already complete because he did not expect physics to explain chemistry, now he expected a quite different completion in the near future, encompassing chemistry as well as physics.

This was still a bit premature. The dream of a final unifying theory really first began to take shape in the mid-1920s, with the discovery of quantum mechanics. This was a new and unfamiliar framework for physics in terms of wave functions and probabilities instead of the particles and forces of Newtonian mechanics. Quantum mechanics suddenly made it possible to calculate the properties not only of individual atoms and their interaction with radiation but also of atoms combined into molecules. It had at last become clear that chemical phenomena are what they are because of the electrical interactions of electrons and atomic nuclei.

This is not to say that college courses in chemistry began to be taught by physics professors or that the American Chemical Society applied for absorption into the American Physical Society. It is difficult enough to use the equations of quantum mechanics to calculate the strength of the binding of two hydrogen atoms in the simplest hydrogen molecule; the special experience and insights of chemists are needed to deal with complicated molecules, especially the very complicated molecules encountered in biology, and the way they react in various circumstances. But the success of quantum mechanics in calculating the properties of very simple molecules made it clear that chemistry works the way it does because of the laws of physics. Paul Dirac, one of the founders of the new quantum mechanics, announced triumphantly in 1929 that "the underlying physical laws necessary for the mathematical theory of a larger part of physics and the whole of chemistry are thus completely known, and the difficulty is only that the application of these laws leads to equations much too complicated to be soluble."

Soon thereafter a strange new problem appeared. The first quantum-mechanical calculations of atomic energies had given results in good agreement with experiment. But, when quantum mechanics was applied not only to the electrons in atoms but also to the electric and magnetic fields that they produce, it turned out that the atom had an infinite energy! Other infinities appeared in other calculations, and for four decades this absurd result appeared as the greatest obstacle to progress in physics. In the end the problem of infinities turned out to be not a disaster, but rather one of the best reasons for optimism about progress toward a final theory. When proper care is given to the definition of masses and electric charges and other constants the infinities all cancel, but *only* in theories of certain special kinds. We may thus find ourselves led mathematically to part or all of the final theory, as the only way of avoiding these infinities. Indeed, the esoteric new theory of strings may already have provided the unique way of avoiding infinities when we reconcile relativity (including general relativity, Einstein's theory of gravitation) with quantum mechanics. If so, it will be a large part of any final theory.

I do not mean to suggest that the final theory will be deduced from pure mathematics. After all, why should we believe that either relativity or quantum mechanics is logically inevitable? It seems to me that our best hope is to identify the final theory as one that is so rigid that it cannot be warped into some slightly different theory without introducing logical absurdities like infinite energies.

There is further reason for optimism in the peculiar fact that progress in physics is often guided by judgments that can only be called aesthetic. This is very odd. Why should a physicist's sense that one theory is more beautiful than another be a useful guide in scientific research? There are several possible reasons for this, but one of them is special to elementary particle physics: the beauty in our present theories may be "but a dream" of the kind of beauty that awaits us in the final theory.

In our century it was Albert Einstein who most explicitly pursued the goal of a final theory. As his biographer Abraham Pais puts it, "Einstein is a typical Old Testament figure, with the

Jehovah-type attitude that there is a law and one must find it." The last thirty years of Einstein's life were largely devoted to a search for a so-called unified field theory that would unify James Clerk Maxwell's theory of electromagnetism with the general theory of relativity, Einstein's theory of gravitation. Einstein's attempt was not successful, and with hindsight we can now see that it was misconceived. Not only did Einstein reject quantum mechanics; the scope of his effort was too narrow. Electromagnetism and gravitation happen to be the only fundamental forces that are evident in everyday life (and the only forces that were known when Einstein was a young man), but there are other kinds of force in nature, including the weak and strong nuclear forces. Indeed, the progress that has been made toward unification has been in unifying Maxwell's theory of the electromagnetic force with the theory of the weak nuclear force, not with the theory of gravitation, where the problem of infinities has been much harder to resolve. Nevertheless Einstein's struggle is our struggle today. It is the search for a final theory.

Talk of a final theory seems to enrage some philosophers and scientists. One is likely to be accused of something awful, like reductionism, or even physics imperialism. This is partly a reaction to the various silly things that might be meant by a final theory, as for instance that discovery of a final theory in physics would mark the end of science. Of course a final theory would not end scientific research, not even pure scientific research, nor even pure research in physics. Wonderful phenomena, from turbulence to thought, will still need explanation whatever final theory is discovered. The discovery of a final theory in physics will not necessarily even help very much in making progress in understanding these phenomena (though it may with some). A final theory will be final in only one sense—it will bring to an end a certain sort of science, the ancient search for those principles that cannot be explained in terms of deeper principles.

CHAPTER II

ON A
PIECE OF CHALK

FOOL: . . . *The reason why the seven stars are no more
than seven is a pretty reason.*

LEAR: *Because they are not eight?*

FOOL: *Yes, indeed. Thou wouldst make a good fool.*

William Shakespeare, *King Lear*

Scientists have discovered many peculiar things, and many beautiful things. But perhaps the most beautiful and the most peculiar thing that they have discovered is the pattern of science itself. Our scientific discoveries are not independent isolated facts; one scientific generalization finds its explanation in another, which is itself explained by yet another. By tracing these arrows of explanation back toward their source we have discovered a striking convergent pattern—perhaps the deepest thing we have yet learned about the universe.

Consider a piece of chalk. Chalk is a substance that is familiar to most people (and especially familiar to physicists who talk to each other at blackboards), but I use chalk as an example here because it was the subject of a polemic that is famous in the history of science. In 1868 the British Association held its annual meeting in the large cathedral and county town of Norwich, in the east of England. It was an exciting time for the scientists and scholars gathered in Norwich. Public attention was drawn to science not only because its importance to technology was becoming unmistakable but even more because science was changing the way that people thought about the world and their place in it. Above all, the publication of Darwin's *On the Origin of Species by Means of Natural Selection* nine years earlier had put science squarely in opposition to the dominant religion of the time. Present at that meeting was Thomas Henry Huxley—distinguished anatomist and fierce controversialist, known to his contemporaries as "Darwin's bulldog." As was his usual practice, Huxley took the opportunity to speak to the working men of the town. The title of his lecture was "On a Piece of Chalk."

I like to imagine Huxley standing at the podium actually holding up a piece of chalk, perhaps dug from the chalk formations that underlie Norwich or borrowed from a friendly carpenter or professor. He began by describing how the chalk layer, hundreds of feet deep, extends not only under much of England but also under Europe and the Levant, all the way to central Asia. The chalk is mostly a simple chemical, "carbonate of lime," or in modern terms calcium carbonate, but microscopic examination shows it to consist of countless fossil shells of tiny animals that lived in ancient seas that once covered Europe. Huxley vividly described how over millions of years these little corpses drifted down to the sea bottom to be compressed into chalk, and how caught here and there in the chalk are fossils of larger animals like crocodiles, animals that appear increasingly different from their modern counterparts as we go down to deeper and deeper levels of chalk, and so must have been evolving during the millions of years the chalk was being laid down.

Huxley was trying to convince the working men of Norwich that the world is much older than the six thousand years allowed by scholars of the Bible and that new living species have appeared and evolved since the beginning. These issues are now long settled—no one with any understanding of science would doubt the great age of the earth or the reality of evolution. The point that I want to make here has to do not with any specific items of scientific knowledge but with the way that they are all connected. For this purpose I begin as Huxley did, with a piece of chalk.

Chalk is white. *Why?* One immediate answer is that it is white because it is not any other color. That is an answer that would have pleased Lear's fool, but in fact it is not so far from the truth. Already in Huxley's time it was known that each color of the rainbow is associated with light of a definite wavelength—longer waves for light toward the red end of the spectrum and shorter waves toward the blue or violet. White light was understood to be a jumble of light of many different wavelengths. When light strikes an opaque substance like chalk, only part of it is reflected; the rest is absorbed. A substance that is any definite color, like the greenish blue of many compounds of copper (e.g., the copper aluminum phosphates in turquoise) or the violet of compounds of chromium, has that color because the substance tends to absorb light strongly at certain wavelengths; the color that we see in the light that the substance reflects is the color associated with light of the wavelengths that are *not* strongly absorbed. For the calcium carbonate of which chalk is composed, it happens that light is especially strongly absorbed only at infrared and ultraviolet wavelengths that are invisible anyway. So light reflected from a piece of chalk has pretty much the same distribution of visible wavelengths as the light that illuminates it. This is what produces the sensation of whiteness, whether from clouds or snow or chalk.

Why? Why do some substances strongly absorb visible light at particular wavelengths and others not? The answer turns out to be a matter of the energies of atoms and of light. This began to be understood with the work of Albert Einstein and Niels Bohr in the first two decades of this century. As Einstein first realized in 1905, a ray of light consists of a stream of enormous numbers of particles, later called *photons.* Photons have no mass or electric charge, but each photon has a definite energy, inversely proportional to the light wavelength. Bohr proposed in 1913 that atoms and molecules can exist only in certain definite *states,* stable configurations having certain definite energies. Although atoms are often likened to little solar systems, there is a crucial difference. In the solar system any planet could be given a little more or less energy by moving it a little farther from or closer to the sun, but the states of an atom are *discrete*—we cannot change the energies of atoms except by certain definite amounts. Normally an atom or molecule is in the state of lowest energy. When an atom or molecule absorbs light, it jumps from a state of lower energy to one of higher energy (and vice versa when light is emitted). Taken together, these ideas of Einstein and Bohr tell us that light can be absorbed by an atom or molecule only if the wavelength of the light has one of certain definite values. These are the wavelengths that correspond to photon energies that are just equal to the difference in energy between the normal state of the atom or molecule and one of its states of higher energy. Otherwise energy would not be conserved when the photon is absorbed by the atom or molecule. Typical copper compounds are greenish blue because there is a particular state of the copper atom with an energy 2 volts higher than the energy of the normal state of the atom, and that is exceptionally easy for the atom to jump to by absorption of a photon with an energy of 2 volts.* Such a photon

---

* A volt when used as a unit of energy is defined as the energy given to one electron when pushed through a wire by a 1-volt electric battery. (When used in this way it should more properly be termed an "electron volt," but as is common in physics, I will just call it a volt.) A micron is a millionth of a meter.

has a wavelength of 0.62 microns, corresponding to a reddish orange color, so the absorption of these photons leaves the remaining reflected light greenish blue. (This is not just an awkward way of restating that these compounds are greenish blue; we see the same pattern of atomic energies when we give energy to the copper atom in different ways, as, e.g., with a beam of electrons.) Chalk is white because the molecules of which it is composed do not happen to have any state that is particularly easy to jump to by absorbing photons of any color of visible light.

Why? Why do atoms and molecules come in discrete states, each with a definite energy? Why are these energy values what they are? Why does light come in individual particles, each with an energy inversely proportional to the wavelength of the light? And why are some states of atoms or molecules particularly easy to jump to by absorption of photons? It was not possible to understand these properties of light or atoms or molecules until the development in the mid-1920s of a new framework for physics known as quantum mechanics. The particles in an atom or molecule are described in quantum mechanics by what is called a wave function. A wave function behaves somewhat like a wave of light or sound, but its magnitude (actually its magnitude squared) gives the probability of finding the particles at any given location. Just as air in an organ pipe can vibrate only in certain definite modes of vibration, each with its own wavelength, so also the wave function for the particles in an atom or molecule can appear only in certain modes or quantum states, each with its own energy. When the equations of quantum mechanics are applied to the copper atom, it is found that one of the electrons in a high-energy outer orbit of the atom is loosely bound and can easily be boosted by absorption of visible light up to the next highest orbit. Quantum-mechanical calculations show that the energies of the atom in these two states differ by 2 volts, equal to the energy of a photon of reddish orange light.* On the other hand the molecules of calcium carbonate in a piece of chalk do not happen to have any similarly loose electrons that could absorb photons of any particular wave length. As to photons, their properties are explained by applying the principles of quantum mechanics in a similar way to light itself. It turns out that light, like atoms, can exist only in certain quantum states of definite energy. For instance, reddish orange light with a wavelength of 0.62 microns can exist only in states with energies equal to zero, or 2 volts, or 4 volts, or 6 volts, and so on, which we interpret as states containing zero or one or two or three or more photons, each photon with an energy of just 2 volts.

Why? Why are the quantum-mechanical equations that govern the particles in atoms what they are? Why does matter consist of these particles, the electrons and the atomic nuclei? For that matter, why is there such a thing as light? Most of these things were rather mysterious in the 1920s and 1930s when quantum mechanics was first applied to atoms and light and have only become reasonably well understood in the last fifteen years or so, with the success of what is called the *standard model* of elementary particles and forces. A key precondition for this new understanding was the reconciliation in the 1940s of quantum mechanics with the other great revolution in twentieth-century physics, Einstein's theory of relativity. The principles of relativity and quantum mechanics are almost incompatible with each other and can coexist only in a limited class of theories. In the nonrelativistic quantum mechanics of the 1920s we could imagine almost any kind of force among electrons and nuclei, but as we shall see, this is not so in a rela-

---

* In a metal these outer electrons leave the individual atoms and flow between them, so there is no special tendency for metallic copper to absorb photons of orange light, which is why it is not greenish blue.

tivistic theory: forces between particles can arise only from the exchange of other particles. Furthermore, all these particles are bundles of the energy, or *quanta*, of various sorts of fields. A field like an electric or magnetic field is a sort of stress in space, something like the various sorts of stress that are possible within a solid body, but a field is a stress in space itself. There is one type of field for each species of elementary particle; there is an electron field in the standard model, whose quanta are electrons; there is an electromagnetic field (consisting of electric and magnetic fields), whose quanta are the photons; there is no field for atomic nuclei, or for the particles (known as protons and neutrons) of which the nuclei are composed, but there are fields for various types of particles called quarks, out of which the proton and neutron are composed; and there are a few other fields I need not go into now. The equations of a field theory like the standard model deal not with particles but with fields; the particles appear as manifestations of these fields. The reason that ordinary matter is composed of electrons, protons, and neutrons is simply that all the other massive particles are violently unstable. The standard model qualifies as an explanation because it is not merely what computer hackers call a kludge, an assortment of odds and ends thrown together in whatever way works. Rather, the structure of the standard model is largely fixed once one specifies the menu of fields that it should contain and the general principles (like the principles of relativity and quantum mechanics) that govern their interactions.

Why? Why does the world consist of just these fields: the fields of the quarks, electron, photon, and so on? Why do they have the properties assumed in the standard model? And for that matter, why does nature obey the principles of relativity and quantum mechanics? Sorry—these questions are still unanswered. Commenting on the present status of physics, the Princeton theorist David Gross gave a list of open questions:

"Now that we understand how it works, we are beginning to ask why are there quarks and leptons, why is the pattern of matter replicated in three generations of quarks and leptons, why are all forces due to local gauge symmetries? Why, why, why?" (Terms used in Gross's list of "whys" are explained in later chapters.) It is the hope of answering these questions that makes elementary particle physics so exciting.

The word "why" is notoriously slippery. The philosopher Ernest Nagel lists ten examples of questions in which "why" is used in ten different senses, such as "Why does ice float on water?" "Why did Cassius plot the death of Caesar?" and "Why do human beings have lungs?" Other examples in which "why" is used in yet other senses come immediately to mind, such as, "Why was I born?" Here, my use of "why" is something like its use in the question "Why does ice float on water?" and is not intended to suggest any sense of conscious purpose.

Even so it is a tricky business to say exactly what one is doing when one answers such a question. Fortunately, it is not really necessary. Scientific explanation is a mode of behavior that gives us pleasure, like love or art. The best way to understand the nature of scientific explanation is to experience the peculiar zing that you get when someone (preferably yourself) has succeeded in actually explaining something. I do not mean that scientific explanation can be pursued without any constraints, any more than can love or art. In all three cases there is a standard of truth that needs to be respected, though of course truth takes different meanings in science or love or art. I also do not mean to say that it is not of any interest to try to formulate some general description of how science is done, only that this is not really necessary in the work of science, any more than it is in love or art.

As I have been describing it, scientific explanation clearly has to do with the deduction of one truth from another. But there is more to explanation than deduction, and also less. Merely deducing one statement from another does not necessarily constitute an explanation, as we see clearly in those cases where either statement can be deduced from the other. Einstein inferred the existence of photons in 1905 from the successful theory of heat radiation that had been proposed five years earlier by Max Planck; nineteen years later Satyendra Nath Bose showed that Planck's theory could be deduced from Einstein's theory of photons. Explanation, unlike deduction, carries a unique sense of *direction*. We have an overwhelming sense that the photon theory of light is more fundamental than any statement about heat radiation and is therefore the explanation of the properties of heat radiation. And in the same way, although Newton derived his famous laws of motion in part from the earlier laws of Kepler that describe the motion of planets in the solar system, we say that Newton's laws explain Kepler's, not the other way around.

Talk of more fundamental truths makes philosophers nervous. We can say that the more fundamental truths are those that are in some sense more comprehensive, but about this, too, it is difficult to be precise. But scientists would be in a bad way if they had to limit themselves to notions that had been satisfactorily formulated by philosophers. No working physicist doubts that Newton's laws are more fundamental than Kepler's or that Einstein's theory of photons is more fundamental than Planck's theory of heat radiation.

A scientific explanation can also be something less than a deduction, for we may say that a fact is explained by some principle even though we cannot deduce it from that principle. Using the rules of quantum mechanics we *can* deduce various properties of the simpler atoms and molecules and even estimate the energy levels of complicated molecules like the calcium-carbonate molecules in chalk. The Berkeley chemist Henry Shaefer reports that "when state-of-the-art theoretical methods are intelligently applied to many problems involving molecules as large as naphthalene, the results may be treated in the same way that one treats reliable experiments." But no one actually solves the equations of quantum mechanics to deduce the detailed wave function or the precise energy of really complicated molecules, such as proteins. Nevertheless, we have no doubt that the rules of quantum mechanics "explain" the properties of such molecules. This is partly because we can use quantum mechanics to deduce the detailed properties of simpler systems like hydrogen molecules and also because we have mathematical rules available that would allow us to calculate all the properties of any molecule to any desired precision if we had a large enough computer and enough computer time.

We may even say that something is explained even where we have no assurance that we will ever be able to deduce it. Right now we do not know how to use our standard model of elementary particles to calculate the detailed properties of atomic nuclei, and we are not certain that we will ever know how to do these calculations, even with unlimited computer power at our disposal. (This is because the forces in nuclei are too strong to allow the sort of calculational techniques that work for atoms or molecules.) Nevertheless, we have no doubt that the properties of atomic nuclei are what they are because of the known principles of the standard model. This "because" does not have to do with our ability actually to deduce anything but reflects our view of the order of nature.

Ludwig Wittgenstein, denying even the possibility of explaining any fact on the basis of any other fact, warned that "at the basis of the whole modern view of the world lies the illusion that the so-called laws of nature are the explanations of natural phenomena." Such warnings leave me cold. To tell a physicist that the laws of nature are not explanations of natural phenomena is like telling a tiger stalking prey that all flesh is grass. The fact that we scientists do not know how to state in a way that philosophers would approve what it is that we are doing in searching for scientific explanations does not mean that we are not doing something worthwhile. We could use help from professional philosophers in understanding what it is that we are doing, but with or without their help we shall keep at it.

We could follow a similar chain of "whys" for each physical property of chalk—its brittleness, its density, its resistance to the flow of electricity. But let's try to enter the labyrinth of explanation through a different door—by considering the chemistry of chalk. As Huxley said, chalk is mostly carbonate of lime or, in modern terms, calcium carbonate. Huxley did not say so, but he probably knew that this chemical consists of the elements calcium, carbon, and oxygen, in the fixed proportions (by weight) 40%, 12%, and 48%, respectively.

Why? Why do we find a chemical compound of calcium, carbon, and oxygen with just these proportions, but not many others, with many other proportions? The answer was worked out by chemists in the nineteenth century in terms of a theory of atoms, actually before there was any direct experimental evidence of the existence of atoms. The weights of the calcium, carbon, and oxygen atoms are in the ratios 40:12:16, and a calcium-carbonate molecule consists of one atom of calcium, one atom of carbon, and *three* atoms of oxygen, so the weights of calcium, carbon, and oxygen in calcium carbonate are in the ratios 40:12:48.

Why? Why do the atoms of various elements have the weights that we observe, and why do molecules consist of just certain numbers of atoms of each type? The numbers of atoms of each type in molecules like calcium carbonate were already known in the nineteenth century to be a matter of the electric charges that the atoms in the molecule exchange with each other. In 1897 it was discovered by J. J. Thomson that these electric charges are carried by negatively charged particles called electrons, particles that are much lighter than whole atoms and that flow through wires in ordinary electric currents. One element is distinguished from another solely by the number of electrons in the atom: one for hydrogen, six for carbon, eight for oxygen, twenty for calcium, and so on. When the rules of quantum mechanics are applied to the atoms of which chalk is composed, it is found that calcium and carbon atoms readily give up two and four electrons, respectively, and that oxygen atoms easily pick up two electrons. Thus the three atoms of oxygen in each molecule of calcium carbonate can pick up the six electrons contributed by one atom of calcium and one atom of carbon; there are just enough electrons to go around. It is the electric force produced by this transfer of electrons that holds the molecule together. What about the atomic weights? We have known since the work of Rutherford in 1911 that almost all of the mass or weight of the atom is contained in a small positively charged nucleus around which the electrons revolve. After some confusion, it was finally realized in the 1930s that atomic nuclei consist of two kinds of particles with nearly the same mass: protons, with positive electric charge equal in magnitude to the electron's negative charge, and neutrons, with no charge at all. The hydrogen nucleus is just one proton. The number of protons must equal the number of electrons in order to keep the atom electrically neutral, and neutrons are needed because the strong attraction between protons and neutrons is essential in holding the nucleus together. Neutrons and protons have nearly the same weight and electrons weigh much less, so to a very good approximation the weight of an atom is simply proportional to the total number of protons and neutrons in its nucleus: one (a proton) for hydrogen, twelve for carbon, sixteen for oxygen, and forty for calcium, corresponding to the atomic weights known but not yet understood in Huxley's time.

Why? Why is there a neutron and a proton, one neutral and one charged, both with about the same mass, and much heavier than the electron? Why do they attract each other with such a strong force that they form atomic nuclei about a hundred thousand times smaller than the atoms themselves? We find the explanation again in the details of our present standard model of elementary particles. The lightest quarks are called the *u* and *d* (for "up" and "down") and have charges $+\frac{2}{3}$ and $-\frac{1}{3}$, respectively (in units where the electron's charge is taken as -1); protons consist of two *u*'s and a *d,* and hence have charge $\frac{2}{3} + \frac{2}{3} - \frac{1}{3} = +1$; neutrons consist of one *u* and two *d*'s and hence have charge $\frac{2}{3} - \frac{1}{3} - \frac{1}{3} = 0$. The proton and neutron masses are nearly equal because these masses arise mostly from strong forces that hold the quarks together, and these forces are the same for *u* and *d* quarks. The electron is much lighter because it does not feel these strong forces. All these quarks and electrons are bundles of the energy of various fields, and their properties follow from the properties of these fields.

So here we are again at the standard model. Indeed, *any* questions about the physical and chemical properties of calcium carbonate lead us in much the same way through a chain of whys down to the same point of convergence: to our present quantum-mechanical theory of elementary particles, the standard model. But physics and chemistry are easy. How about something tougher, like biology?

Our piece of chalk is not a perfect crystal of calcium carbonate, but it is also not a disorganized mess of individual molecules like a gas. Rather, as Huxley explained in his talk in Norwich, chalk is composed of the skeletons of tiny animals who absorbed calcium salts and carbon dioxide from ancient seas and used these chemicals as raw materials to build little shells of calcium carbonate around their soft bodies. It takes no imagination to see why this was to their advantage—the sea is not a safe place for unprotected morsels of protein. But this does not in itself explain why plants and animals develop organs like calcium carbonate shells that help them survive; needing is not the same as getting. The key was provided by the work of Darwin and Wallace that Huxley did so much to popularize and defend. Living things display inheritable variations—some helpful and some not—but it is the organisms that happen to carry helpful variations that tend to survive and pass these characteristics on to their offspring. But why are there variations, and why are they inheritable? This was finally explained in the 1950s in terms of the structure of a very large molecule, DNA, that serves as a template for assembling proteins out of amino acids. The DNA molecule forms a double helix that stores genetic information in a code based on the sequence of chemical units along the two strands of the helix. Genetic information is propagated when the double helix splits and each of its two strands assembles a copy of itself; inheritable variations occur when some accident disturbs the chemical units making up the strands of the helix.

Once down to the level of chemistry, the rest is relatively easy. True, DNA is too complicated to allow us to use the equations of quantum mechanics to work out its structure. But the structure is understood well enough through the ordinary rules of chemistry, and no one doubts that with a large enough computer we could in principle explain all the properties of DNA by solving the equations of quantum mechanics for electrons and the nuclei of a few common elements, whose properties are explained in turn by the standard model. So again we find ourselves at the same point of convergence of our arrows of explanation.

I have papered over an important difference between biology and the physical sciences: the element of history. If by "chalk" we mean "the material of the white cliffs of Dover" or "the thing in Huxley's hand," then the statement that chalk is 40% calcium, 12% carbon, and 48% oxygen must find its explanation in a mixture of the universal and the historical, including accidents that occurred in the history of our planet or in the life of Thomas Huxley. The propositions that we can hope to explain in terms of the final laws of nature are those about universals. One such universal is the statement that (at sufficiently low temperatures and pressures) there exists a chemical compound consisting of precisely these proportions of calcium, carbon, and oxygen. We think that such statements are true everywhere in the universe and throughout all time. In the same way, we can make universal propositions about the properties of DNA, but the fact that there are living creatures on earth that use DNA to pass on random variations from one generation to the next depends on certain historical accidents: there is a planet like the earth; life and genetics somehow got started; and a long time has been available for evolution to do its work.

Biology is not unique in involving this element of history. The same is true of many other sciences, such as geology and astronomy. Suppose we pick up our piece of chalk one more time and ask why there is enough calcium, carbon, and oxygen here on earth to provide raw materials for the fossil shells that make up chalk? That is easy—these elements are common throughout most of the universe. But why is that? Again, we must appeal to a blend of history and universal principles. Using the standard model of elementary particles, we know how to follow the course of nuclear reactions in the standard "big-bang" theory of the universe well enough to be able to calculate that the matter formed in the first few minutes of the universe was about three-quarters hydrogen and one-quarter helium, with only a trace of other elements, chiefly very light ones like lithium. This is the raw material out of which heavier elements were later formed in stars. Calculations of the subsequent course of nuclear reactions in stars show that the elements that are most abundantly produced are those whose nuclei are most tightly bound, and these elements include carbon, oxygen, and calcium. The stars dump this material into the interstellar medium in various ways, in stellar winds and supernova explosions, and it is out of this medium, rich in the constituents of chalk, that second-generation stars like the sun and their planets were formed. But this scenario still depends on a historical assumption—that there was a more-or-less homogeneous big bang, with about ten billion photons for every quark. Efforts are being made to explain this assumption in various speculative cosmological theories, but these theories rest in turn on other historical assumptions.

It is not clear whether the universal and the historical elements in our sciences will remain forever distinct. In modern quantum mechanics as well as in Newtonian mechanics there is a clear separation between the conditions that tell us the initial state of a system (whether the system is the whole universe, or just a part of it), and the laws that govern its subsequent evolution. But it is possible that eventually the initial conditions will appear as part of the laws of nature. One simple example of how this is possible is provided by what is called the steady-state cosmology, proposed in the late 1940s by Herman Bondi and Thomas Gold and (in a rather different version) by Fred Hoyle. In this picture, although the galaxies are all rushing apart from each other (a fact often expressed in the somewhat misleading statement that the universe is expanding*), new

---

* It is misleading to say that the universe is expanding, because solar systems and galaxies are not expanding, and space itself is not expanding. The galaxies are rushing apart in the way that any cloud of particles will rush apart once they are set in motion away from each other.

matter is continually being created to fill up the expanding intergalactic voids, at a rate that just manages to keep the universe looking always the same. We do not have a believable theory of how this continual creation of matter might take place, but it is plausible that if we did have such a theory we might be able to use it to show that the expansion of the universe tends to an equilibrium rate at which creation just balances expansion, like the way that prices are supposed to adjust themselves until supply equals demand. In such a steady-state theory there are no initial conditions because there is no beginning, and instead we can deduce the appearance of the universe from the condition that it does not change.

The original version of the steady-state cosmology has been pretty well ruled out by various astronomical observations, chief among them the discovery in 1964 of microwave radiation that seems to be left over from a time when the universe was much hotter and denser. It is possible that the steady-state idea may be revived on a grander scale, in some future cosmological theory in which the present expansion of the universe appears as merely a fluctuation in an eternal but constantly fluctuating universe that on average is always the same. There are also more subtle ways that the initial conditions might perhaps some day be deduced from the final laws. James Hartle and Stephen Hawking have proposed one way that this fusion of physics and history might be found in the application of quantum mechanics to the whole universe. Quantum cosmology is right now a matter of active controversy among theorists; the conceptual and mathematical problems are very difficult, and we do not seem to be moving toward any definite conclusions.

In any case, even if the initial conditions of the universe can ultimately be incorporated in or deduced from the laws of nature, as a practical matter we will never be able to eliminate the accidental and historical elements of sciences like biology and astronomy and geology. Stephen Gould has used the weird fossils of the Burgess Shale in British Columbia to illustrate how little inevitability there is in the pattern of biological evolution on earth. Even a very simple system can exhibit a phenomenon known as *chaos* that defeats our efforts to predict the system's future. A chaotic system is one in which nearly identical initial conditions can lead after a while to entirely different outcomes. The possibility of chaos in simple systems has actually been known since the beginning of the century; the mathematician and physicist Henri Poincaré showed then that chaos can develop even in a system as simple as a solar system with only two planets. The dark gaps in the rings of Saturn have been understood for many years to occur at just those positions in the rings from which any orbiting particles would be ejected by their chaotic motion. What is new and exciting about the study of chaos is not the discovery that chaos exists but that certain kinds of chaos exhibit some nearly universal properties that can be analyzed mathematically.

The existence of chaos does not mean that the behavior of a system like Saturn's rings is somehow not completely determined by the laws of motion and gravitation and its initial conditions but only that as a practical matter we can not calculate how some things (such as particle orbits in the dark gaps in Saturn's rings) evolve. To put this a little more precisely: the presence of chaos in a system means that for any given accuracy with which we specify the initial conditions, there will eventually come a time at which we lose all ability to predict how the system will behave, but it is still true that however far into the future we want to be able to predict the behavior of a physical system governed by Newton's laws, there is some degree of accuracy with which a measurement of the initial conditions would allow us to make this prediction. (It is like saying that, although any automobile that keeps going will eventually run out of gasoline no matter how much we put in the tank, still no matter how far we want to go there is always some amount of gasoline that would get us there.) In other words, the discovery of chaos did not abolish the determinism of prequantum physics, but it did force us to be a bit more careful in saying what we mean by this determinism. Quantum mechanics is not deterministic in the same sense as Newtonian mechanics; Heisenberg's uncertainty principle warns that we cannot measure the position and velocity of a particle precisely at the same time, and, even if we make all of the measurements that are possible at one time, we can predict only probabilities about the results of experiments at any later time. Nevertheless we shall see that even in quantum mechanics there is still a sense in which the behavior of any physical system is completely determined by its initial conditions and the laws of nature.

Of course whatever determinism survives in principle does not help us very much when we have to deal with real systems that are not simple, like the stock market or life on earth. The intrusion of historical accidents sets permanent limits on what we can ever hope to explain. Any explanation of the present forms of life on earth must take into account the extinction of the dinosaurs sixty-five million years ago, which is currently explained by the impact of a comet, but no one will ever be able to explain why a comet happened to hit the earth at just that time. The most extreme hope for science is that we will be able to trace the explanations of all natural phenomena to final laws *and* historical accidents.

The intrusion of historical accidents into science means also that we have to be careful what sort of explanations we demand from our final laws. For instance, when Newton first proposed his laws of motion and gravitation the objection was raised that these laws did not explain one of the outstanding regularities of the solar system, that all the planets are going around the sun in the same direction. Today we understand that this is a matter of history. The way that the planets revolve around the sun is a consequence of the particular way that the solar system condensed out of a rotating disk of gas. We would not expect to be able to deduce it from the laws of motion and gravitation alone. The separation of law and history is a delicate business, one we are continually learning how to do as we go along.

Not only is it possible that what we now regard as arbitrary initial conditions may ultimately be deduced from universal laws—it is also conversely possible that principles that we *now* regard as universal laws will eventually turn out to represent historical accidents. Recently a number of theoretical physicists have been playing with the idea that what we usually call the universe, the expanding cloud of galaxies that extends in all directions for at least tens of billions of light years, is merely a subuniverse, a small part of a much larger megauniverse consisting of many such parts, in each of which what we call the constants of nature (the electric charge of the electron, the ratios of elementary particle masses, and so on) may take different values. Perhaps even what we now call the laws of nature will be found to vary from one subuniverse to another. In that case, the explanation for the constants and laws that we have discovered may involve an irreducible historical element: the accident that we are in the particular subuniverse we inhabit. But, even if there turns out to be something in these ideas, I do not think that we will have to give up our dreams of discovering final laws of nature; the final laws would be megalaws that determine the probabilities of being in different types of subuniverse. Sidney Coleman and others have already made brave steps toward calculating these probabilities by applying quantum mechanics to the whole megauniverse. I should stress that these are very speculative ideas, not fully formulated mathematically, and so far without experimental support.

I have so far confessed to two problems in the notion of chains of explanation that lead down to final laws: the intrusion of historical accidents and the complexity that prevents our being actually able to explain everything even when we consider only universals, free of the element of history. There is one other problem that must be confronted, one associated with the buzz-word "emergence." As we look at nature at levels of greater and greater complexity, we see phenomena emerging that have no counterpart at the simpler levels, least of all at the level of the elementary particles. For instance, there is nothing like intelligence on the level of individual living cells, and nothing like life on the level of atoms and molecules. The idea of emergence was well captured by the physicist Philip Anderson in the title of a 1972 article: "More Is Different." The emergence of new phenomena at high levels of complexity is most obvious in biology and the behavioral sciences, but it is important to recognize that such emergence does not represent something special about life or human affairs; it also happens within physics itself.

The example of emergence that has been historically most important in physics is thermodynamics, the science of heat. As originally formulated in the nineteenth century by Carnot, Clausius, and others, thermodynamics was an autonomous science, not deduced from the mechanics of particles and forces but built on concepts like entropy and temperature that have no counterparts in mechanics. Only the first law of thermodynamics, the conservation of energy, provided a bridge between mechanics and thermodynamics. The central principle of thermodynamics was the second law, according to which (in one formulation) physical systems possess not only an energy and a temperature but also a certain quantity called entropy, which always increases with time in any closed system and reaches a maximum when the system is in equilibrium. This is the principle that forbids the Pacific Ocean from spontaneously transferring so much heat energy to the Atlantic that the Pacific freezes and the Atlantic boils; such a cataclysm need not violate the conservation of energy, but it is forbidden because it would decrease the entropy.

Nineteenth-century physicists generally took the second law of thermodynamics as an axiom, derived from experience, as fundamental as any other law of nature. At the time this was not unreasonable. Thermodynamics was seen to work in vastly different contexts, from the behavior of steam (the problem that gave thermodynamics its start) to freezing and boiling and chemical reactions. (Today we would add more exotic examples; astronomers have discovered that the clouds of stars in globular clusters in our own and other galaxies behave like gases with definite temperatures, and the work of Jacob Bekenstein and Hawking has shown theoretically that a black hole has an entropy proportional to its surface area.) If thermodynamics is this universal, how can it be logically related to the physics of specific types of particles and forces?

Then in the second half of the nineteenth century the work of a new generation of theoretical physicists (including Maxwell in Scotland, Ludwig Boltzmann in Germany, and Josiah Willard Gibbs in America) showed that the principles of thermodynamics could in fact be deduced mathematically, by an analysis of the probabilities of different configurations of certain kinds of system, those systems whose energy is shared among a very large number of subsystems, as for instance a gas whose energy is shared among the molecules of which it is composed. (Ernest Nagel gave this as a paradigmatic example of the reduction of one theory to another.) In this statistical mechanics, the heat energy of a gas is just the kinetic energy of its particles; the entropy is a measure of the disorder of the system; and the second law of thermodynamics expresses the tendency of isolated systems to become more disorderly. The flow of all the heat of the oceans into the Atlantic would represent an increase of order, which is why it does not happen.

For a while during the 1880s and 1890s a battle was fought between the supporters of the new statistical mechanics and those like Planck and the chemist Wilhelm Ostwald who continued to maintain the logical independence of thermodynamics. Ernst Zermelo went even further and argued that, because on the basis of statistical mechanics the decrease of entropy would be very unlikely but not impossible, the assumptions about molecules on which statistical mechanics is based must be wrong. This battle was won by statistical mechanics, after the reality of atoms and molecules became generally accepted early in this century. Nevertheless, even though thermodynamics has been explained in terms of particles and forces, it continues to deal with emergent concepts like temperature and entropy that lose all meaning on the level of individual particles.

Thermodynamics is more like a mode of reasoning than a body of universal physical law; wherever it applies it always allows us to justify the use of the same principles, but the explanation of why thermodynamics does apply to any particular system takes the form of a deduction using the methods of statistical mechanics from the details of what the system contains, and this inevitably leads us down to the level of the elementary particles. In terms of the image of arrows of explanation that I invoked earlier, we can think of thermodynamics as a certain pattern of arrows that occurs again and again in very different physical contexts, but, wherever this pattern of explanation occurs, the arrows can be traced back by the methods of statistical mechanics to deeper laws and ultimately to the principles of elementary particle physics. As this example shows, the fact that a scientific theory finds applications to a wide variety of different phenomena does not imply anything about the autonomy of this theory from deeper physical laws.

The same maxim applies to other areas of physics, such as the related topics of chaos and turbulence. Physicists working in these areas have found certain patterns of behavior that occur again and again in very different contexts; for instance, there is thought to be a universality of a sort in the distribution of energy in eddies of various size in turbulent fluids of all sorts, from the turbulence of the tidal flow in Puget Sound to the turbulence in the interstellar gas produced by a passing star. But not all fluid flows are turbulent, and turbulence when it occurs does not always exhibit these "universal" properties. Whatever the mathematical reasoning that accounts for the universal properties of turbulence, we still have to explain *why* this reasoning should apply to any particular turbulent fluid, and this question inevitably will be answered in terms of accidents (the speed of the tidal flow and the shape of the channel) and universals (the laws of fluid motion and the properties of water) that in turn must be explained in terms of deeper laws.

Similar remarks apply to biology. Here most of what we see depends on historical accidents, but there are some roughly universal patterns, like the rule of population biology that dictates that males and females tend to be born in equal numbers. (In 1930 the geneticist Ronald Fisher explained that once a species develops a tendency to produce, say, more males than females, any gene that gives individuals a tendency to produce more females than males spreads through the population, because the female offspring of individuals carrying this gene encounter less competition in finding a mate.) Rules like this apply to a wide variety of species and might be expected to apply even to life discovered on other planets if it reproduced sexually. The reasoning that leads to these rules is the same whether it is applied to humans or birds or extraterrestrials, but the reasoning always rests on certain assumptions about the organisms involved, and, if we ask *why* these assumptions should be found to be correct, we must seek the answer partly in terms of historical accidents and partly in terms of universals like the properties of DNA (or whatever takes its place on other planets) that must in turn find their explanation in physics and chemistry, and hence in the standard model of elementary particles.

This point tends to get obscured because, in the actual work of thermodynamics or fluid dynamics or population biology, scientists use languages that are special to their own fields, speaking of entropy or eddies or reproductive strategies and not the language of elementary particles. This is not only because we are unable to use our first principles actually to calculate complicated phenomena; it is also a reflection of the sort of question we want to ask about these phenomena. Even if we had an enormous computer that could follow the history of every elementary particle in a tidal flow or a fruit fly, this mountain of computer printout would not be of much use to someone who wanted to know whether the water was turbulent or the fly was alive.

There is no reason to suppose that the convergence of scientific explanations must lead to a convergence of scientific methods. Thermodynamics and chaos and population biology will each continue to operate with its own language, under its own rules, whatever we learn about the elementary particles. As the chemist Roald Hoffman says, "Most of the useful concepts of chemistry . . . are imprecise. When reduced to physics, they tend to disappear." In an attack on those who seek to reduce chemistry to physics, Hans Primas listed some of the useful concepts of chemistry that were in danger of being lost in this reduction: valence, bond structure, localized orbitals, aromaticity, acidity, color, smell, and water repellency. I see no reason why chemists should stop speaking of such things as long as they find it useful or interesting. But the fact that they continue to do so does not cast doubt on the fact that all these notions of chemistry work the way they do because of the underlying quantum mechanics of electrons, protons, and neutrons. As Linus Pauling puts it, "There is no part of chemistry that does not depend, in its fundamental theory, upon quantum principles."

Of all the areas of experience that we try to link to the principles of physics by arrows of explanation, it is consciousness that presents us with the greatest difficulty. We know about our own conscious thoughts directly, without the intervention of the senses, so how can consciousness ever be brought into the ambit of physics and chemistry? The physicist Brian Pippard, who held Maxwell's old chair as Cavendish Professor at the University of Cambridge, has put it thus: "What is surely impossible is that a theoretical physicist, given unlimited computing power, should deduce from the laws of physics that a certain complex structure is aware of its own existence."

I have to confess that I find this issue terribly difficult, and I have no special expertise on such matters. But I think I disagree with Pippard and the many others who take the same position. It is clear that there is what a literary critic might call an objective correlative to consciousness; there are physical and chemical changes in my brain and body that I observe to be correlated (either as cause or effect) with changes in my conscious thoughts. I tend to smile when pleased; my brain shows different electrical activity when I am awake or asleep; powerful emotions are triggered by hormones in my blood; and I sometimes speak my thoughts. These are not consciousness itself; I can never express in terms of smiles or brain waves or hormones or words what it *feels* like to be happy or sad. But setting consciousness to one side for a moment, it seems reasonable to suppose that these objective correlatives to consciousness can be studied by the methods of science and will eventually be explained in terms of the physics and chemistry of the brain and body. (By "explained" I do not necessarily mean that we will be able to predict everything or even very much, but that we will understand why smiles and brain waves and hormones work the way they do, in the same sense that, although we cannot predict next month's weather, still we understand why the weather works the way it does.)

In Pippard's own Cambridge there is a group of biologists headed by Sydney Brenner who have completely worked out the wiring diagram of the nervous system of a small nematode worm, C. elegans, so that they already have a basis for understanding in some sense everything about why that worm behaves the way it does. (What is lacking so far is a program based on this wiring diagram that can generate the worm's observed behavior.) Of course a worm is not a human. But between a worm and a human there is a continuum of animals with increasingly complex nervous systems, spanning insects and fishes and mice and apes. Where is one to draw the line?

Suppose then that we will come to understand the objective correlatives to consciousness in terms of physics (including chemistry) and that we will also understand how they evolved to be what they are. It is not unreasonable to hope that when the objective correlatives to consciousness have been explained, somewhere in our explanations we shall be able to recognize something, some physical system for processing information, that corresponds to our experience of consciousness itself, to what Gilbert Ryle has called "the ghost in the machine." That may not be an explanation of consciousness, but it will be pretty close.

There is no guarantee that progress in other fields of science will be assisted directly by anything new that is discovered about the elementary particles. But (I repeat, and not for the last time) I am concerned here not so much with what scientists do, because this inevitably reflects both human limitations and human interests, as I am with the logical order built into nature itself. It is in this sense that branches of physics like thermodynamics and other sciences like chemistry and biology may be said to rest on deeper laws, and in particular on the laws of elementary particle physics.

In speaking here of a logical order of nature I have been tacitly taking what a historian of philosophy would call a "realist" position—realist not in the everyday modern sense of being hardheaded and without illusions, but in a much older sense, of believing in the reality of abstract ideas. A medieval realist believed in the reality of universals like Plato's forms, in opposition to nominalists like William of Ockham, who declared them to be mere names. (My use of the word "realist" would have pleased one of my favorite authors, the Victorian George Gissing, who wished that "the words *realism* and *realist* might never again be used, save in their proper sense by writers on scholastic philosophy.") I certainly do not want to enter this debate on the side of Plato. My argument here is for the reality of the laws of nature, in opposition to the modern positivists, who accept the reality only of that which can be directly observed.

When we say that a thing is real we are simply expressing a sort of respect. We mean that the thing must be taken seriously because it can affect us in ways that are not entirely in our control and because we cannot learn about it without making an effort that goes beyond our own imagination. This much is true for instance of the chair on which I sit (to take a favorite example of philosophers) and does not so much constitute evidence that the chair is real but is rather just what we *mean* when we say that the chair is real. As a physicist I perceive scientific explanations and laws as things that are what they are and cannot be made up as I go along, so my relation to these laws is not so different from my relation to my chair, and I therefore accord the laws of nature (to which our present laws are an approximation) the honor of being real. This impression is reinforced when it turns out that some law of nature is not what we thought it was, an experience similar to finding that a chair is not in place when one sits down. But I have to admit that my willingness to grant the title of "real" is a little like Lloyd George's willingness to grant titles of nobility; it is a measure of how little difference I think the title makes.

This discussion of the reality of the laws of nature might become less academic if we made contact with other intelligent beings on distant planets who had also worked out scientific explanations for natural phenomena. Would we find that they had discovered the same laws of nature? Whatever laws were discovered by extraterrestrials would naturally be expressed in different language and notation, but we could still ask whether there is some sort of correspondence between their laws and our laws. If so, it would be hard to deny the objective reality of these laws.

Of course we do not know what the answer would be, but here on earth we have already seen a small scale test of a similar question. What we call modern physical science happened to get started in Europe at the end of the sixteenth century. Those who doubt the reality of the laws of nature might have guessed that just as other parts of the world have kept their own languages and religions, so they would also have kept their own scientific traditions, eventually developing laws of physical science completely different from those of Europe. Of course that did not happen: the physics of modern Japan and India is the same as that of Europe and America. I admit that this argument is not entirely convincing, because the whole world has been profoundly influenced by other aspects of Western civilization, from military organization to blue jeans. Nevertheless, the experience of listening to a discussion of quantum field theory or weak interactions in a seminar room in Tsukuba or Bombay gives me a powerful impression that the laws of physics have an existence of their own.

Our discovery of the connected convergent pattern of scientific explanation has profound implications, and not just for scientists. Alongside the main stream of scientific knowledge there are isolated little pools of what (to choose a neutral term) I might call would-be sciences: astrology, precognition, "channeling," clairvoyance, telekinesis, creationism, and their kin. If it could be shown that there is any truth to any of these notions it would be the discovery of the century, much more exciting and important than anything going on today in the normal work of physics. So what should a thoughtful citizen conclude when it is claimed by a professor or a film star or Time-Life Books that there is evidence for the validity of one of the would-be sciences?

Now, the conventional answer would be that this evidence must be tested with an open mind and without theoretical preconceptions. I do not think that this is a useful answer, but this view seems to be widespread. Once in a television interview I said that in believing in astrology one would be turning one's back on all of modern science. I then received a polite letter from a former chemist and metallurgist in New Jersey who took me to task because I had not personally studied the evidence for astrology. Similarly, when Philip Anderson recently wrote disparagingly of belief in clairvoyance and telekinesis, he was upbraided by a Princeton colleague, Robert Jahn, who was experimenting with what Jahn calls "consciousness-related anomalous phenomena." Jahn complained that "although his [Anderson's] office is only a few hundred yards from my own, he has not visited our laboratory, discussed any of his concerns with me directly or apparently even read with care any of our technical literature."

What Jahn and the New Jersey chemist and others who agree with them are missing is the sense of the connectedness of scientific knowledge. We do not understand everything, but we understand enough to know that there is no room in our world for telekinesis or astrology. What possible physical signal from our brains could move distant objects and yet have no effect on any scientific instruments? Defenders of astrology sometimes point to the undoubted effects of the moon and sun in producing tides, but the effects of the gravitational fields of the other planets are much too small to have detectable effects on the earth's oceans, much less on anything as small as a person. (I will not belabor the point, but similar remarks apply to any effort to explain clairvoyance or precognition or the other would-be sciences in terms of standard science.) In any case, the correlations predicted by astrologers are not of the sort that might arise from some very subtle gravitational effect; the astrologers do not claim merely that a certain configuration of planets affects life here on earth, but that these effects differ for each person according to the date and hour of his birth! In fact, I do not think that most people who believe in astrology imagine that it works the way it does because of gravitation or any other agency within the scope of physics; I think they believe that astrology is an autonomous science, with its own fundamental laws, not to be explained in terms of physics or anything else. One of the great services provided by the discovery of the pattern of scientific explanation is to show us that there are no such autonomous sciences.

But still, should we not test astrology and telekinesis and the rest to make sure that there is nothing to them? I have nothing against anyone testing anything they want to, but I do want to explain why I would not bother to do so myself and would not recommend the task to anyone else. At any one moment one is presented with a wide variety of innovative ideas that might be followed up: not only astrology and such, but many ideas much closer to the main stream of science, and others that are squarely within the scope of modern scientific research. It does no good to say that *all* these ideas must be thoroughly tested; there is simply no time. I receive in the mail every week about fifty preprints of articles on elementary particle physics and astrophysics, along with a few articles and letters on all sorts of would-be science. Even if I dropped everything else in my life, I could not begin to give all of these ideas a fair hearing. So what am I to do? Not only scientists but everyone else faces a similar problem. For all of us, there is simply no alternative to making a judgment as well as we can that some of these ideas (perhaps most of them) are not worth pursuing. And our greatest aid in making this judgment is our understanding of the pattern of scientific explanation.

When the Spanish settlers in Mexico began in the sixteenth century to push northward into the country known as Texas, they were led on by rumors of cities of gold, the seven cities of Cibola. At the time that was not so unreasonable. Few Europeans had been to Texas, and for all anyone knew it might contain any number of wonders. But suppose that someone today reported evidence that there are seven golden cities somewhere in modern Texas. Would you open-mindedly recommend mounting an expedition to search every corner of the state between the Red River and the Rio Grande to look for these cities? I think you would make the judgment that we already know so much about Texas, so much of it has been explored and settled, that it is simply not worthwhile to look for mysterious golden cities. In the same way, our discovery of the connected and convergent pattern of scientific explanations has done the very great service of teaching us that there is no room in nature for astrology or telekinesis or creationism or other superstitions.

CHAPTER III

TWO CHEERS
FOR REDUCTIONISM

Darling, you and I know the reason why
The summer sky is blue,
And we know why birds in the trees
Sing melodies too.

Meredith Willson, *You and I*

If you go around asking why things are the way they are, and if, when you are given an explanation in terms of some scientific principle, you ask why that principle is true, and if like an ill-mannered child you persist in asking why? why? why? then sooner or later someone is going to call you a reductionist. Different people mean different things by this, but I suppose that one common feature of everyone's idea of reductionism is a sense of hierarchy, that some truths are less fundamental than others to which they may be reduced, as chemistry may be reduced to physics. Reductionism has become a standard Bad Thing in the politics of science; Canada's Science Council recently attacked the Canadian Agricultural Services Coordinating Committee for being dominated by reductionists. (Presumably the Science Council meant that the Coordinating Committee put too much emphasis on plant biology and chemistry.) Elementary particle physicists are particularly liable to be called reductionists, and the dislike of reductionism has often soured relationships between them and other scientists.

The opponents to reductionism occupy a wide ideological spectrum. At its most reasonable end are those who object to the more naive forms of reductionism. I share their objections. I consider myself a reductionist, but I do not think that the problems of elementary particle physics are the only interesting and profound ones in science, or even in physics. I do not think that chemists should drop everything else they are doing and devote themselves to solving the equations of quantum mechanics for various molecules. I do not think that biologists should stop thinking about whole plants and animals and think only about cells and DNA. For me, reductionism is not a guideline for research programs, but an attitude toward nature itself. It is nothing more or less than the perception that scientific principles are the way they are because of deeper scientific principles (and, in some cases, historical accidents) and that all these principles can be traced to one simple connected set of laws. At this moment in the history of science it appears that the best way to approach these laws is through the physics of elementary particles, but that is an incidental aspect of reductionism and may change.

At the other end of the spectrum are the opponents of reductionism who are appalled by what they feel to be the bleakness of modern science. To whatever extent they and their world can be reduced to a matter of particles or fields and their interactions, they feel diminished by that knowledge. Dostoevsky's underground man imagines a scientist telling him, "Nature doesn't consult you; it doesn't give a damn for your wishes or whether its laws please you or do not please you. You must accept it as it is. . . ." and he replies, "Good God, what do I care about the laws of nature and arithmetic if for one reason or another, I don't like these laws. . . ." At its nuttiest extreme are those with holistics in their heads, those whose reaction to reductionism takes the form of a belief in psychic energies, life forces that cannot be described in terms of the ordinary laws of inanimate nature. I would not try to answer these critics with a pep talk about the beauties of modern science. The reductionist worldview *is* chilling and impersonal. It has to be accepted as it is, not because we like it, but because that is the way the world works.

In the middle of the spectrum of antireductionists there is a group that is less disinterested and far more important. They are the scientists who are infuriated to hear it said that their branches of science rest on the deeper laws of elementary particle physics.

For some years I have been wrangling about reductionism with a good friend, the evolutionary biologist Ernst Mayr, who among other things gave us our best working definition of a biological species. It started when, in a 1985 article, he pounced on a line in a *Scientific American* article (on other matters) that I had written in 1974. In this article I had mentioned that in physics we hope to find a few simple general laws that would explain why nature is the way it is and that at present the closest we can come to a unified view of nature is a description of elementary particles and their mutual interactions. Mayr in his article called this "a horrible example of the way physicists think" and referred to me as "an uncompromising reductionist." I responded in an article in *Nature* that I am not an uncompromising reductionist; I am a compromising reductionist.

There followed a frustrating correspondence, in which Mayr outlined a classification of different sorts of reductionism and identified my particular version of this heresy. I did not

understand the classification; all his categories sounded alike to me, and none of them described my own views. He in turn (it seemed to me) did not understand the distinction that *I* was making, between reductionism as a general prescription for progress in science, which is not my view, and reductionism as a statement of the order of nature, which I think is simply true.*
Mayr and I are still on good terms but have given up on converting each other.

Most serious for the nation's research planning has been the opposition to reductionism within physics itself. The reductionist claims of elementary particle physics are deeply annoying to some physicists who work in other fields, such as condensed matter physics, and who feel themselves in competition for funds with the elementary particle physicists. These arguments have been raised to new levels of ill feeling by the proposal to spend billions of dollars on a particle accelerator, the Superconducting Super Collider. In 1987 the executive director of the American Physical Society's office of public affairs remarked that the Super Collider project "is perhaps the most divisive issue ever to confront the physics community." During the time

---

* As far as I can understand it, Mayr distinguishes three kinds of reductionism: *constitutive reductionism* (or ontological reductionism, or analysis), which is a method of studying objects by inquiring into their basic constituents; *theory reductionism,* which is the explanation of a whole theory in terms of a more inclusive theory; and *explanatory reductionism,* which is the view "that the mere knowledge of its ultimate components would be sufficient to explain a complex system." The main reason I reject this categorization is that none of these categories has much to do with what I am talking about (though I suppose theory reductionism comes closest). Each of these three categories is defined by what scientists actually do or have done or could do; I am talking about nature itself. For instance, even though physicists cannot actually explain the properties of very complicated molecules like DNA in terms of the quantum mechanics of electrons, nuclei, and electric forces, and even though chemistry survives to deal with such problems with its own language and concepts, still there are no autonomous principles of chemistry that are simply independent truths, not resting on deeper principles of physics.

that I served on the board of overseers of the Super Collider project I and the other members of the board had to do a good deal of public explaining about the aims of the project. One of the members of the board argued that we should not give the impression that we think that elementary particle physics is more fundamental than other fields, because it just tended to enrage our friends in other areas of physics.

The reason we give the impression that we think that elementary particle physics is more fundamental than other branches of physics is because it is. I do not know how to defend the amounts being spent on particle physics without being frank about this. But by elementary particle physics being more fundamental I do not mean that it is more mathematically profound or that it is more needed for progress in other fields or anything else but only that it is closer to the point of convergence of all our arrows of explanation.

Chief among the physicists who are unhappy about the pretensions of particle physics is Philip Anderson of Bell Labs and Princeton, a theoretical physicist who has provided many of the most pervasive ideas underlying modern condensed matter physics (the physics of semiconductors and superconductors and such). Anderson testified against the Super Collider project in the same congressional committee hearings at which I testified in 1987. He felt (and I do, too) that research in condensed matter physics is underfunded by the National Science Foundation. He felt (and I do, too) that many graduate students are seduced by the glamour of elementary particle physics, when they could have careers that would be more scientifically satisfying in condensed matter physics and allied fields. But Anderson went on to claim that ". . . they [the results of particle physics] are in no sense more fundamental than what Alan Turing did in founding the computer science, or what Francis Crick and James Watson did in discovering the secret of life."

In no sense more fundamental? This is the essential point where Anderson and I part company. I will pass over the work of Turing and the beginnings of computer science, which seem to me to belong more to mathematics or technology than to the usual framework of natural science. Mathematics itself is never the explanation of anything—it is only the means by which we use one set of facts to explain another, and the language in which we express our explanations. But Anderson's description of Crick and Watson's discovery of the double-helix structure of the DNA molecule (which provides the mechanism by which genetic information is preserved and transmitted) as the secret of life furnishes ammunition for *my* argument. This description of the DNA discovery would strike some biologists as just as wrongheadedly reductionist as the claims of particle physicists seem to Anderson. For instance, Harry Rubin wrote a few years ago that "[the] DNA revolution led a generation of biologists to believe that the secret of life lay entirely in the structure and function of DNA. This faith is misplaced and the reductionist programme must be supplemented with a new conceptual framework." My friend Ernst Mayr has been fighting for years against the reductionist trend in biology that he fears aims at reducing all we know about life to the study of DNA, and argues that "to be sure the chemical nature of a number of black boxes in the classical genetic theory were filled in by the discovery of DNA, RNA, and others, but this did not affect in any way the nature of transmission genetics."

I am not going to enter this debate among biologists, least of all on the side of the antireductionists. There is no doubt that DNA has been of enormous importance in many areas of biology. Still, there are *some* biologists whose work is not directly affected by discoveries in molecular biology. The knowledge of the structure of DNA is of little help to a population ecologist trying to explain the diversity of plant species in tropical rain forests or perhaps to a biomechanician trying to understand the flight of butterflies. My point is that, even if no biologist received any help in his or her work from the discoveries of molecular biology, there would still be an important sense in which Anderson would have been right in talking about the secret of life. It is not that the *discovery* of DNA was fundamental to all of the *science* of life, but rather that DNA itself is fundamental to all life itself. Living things are the way they are because through natural selection they have evolved to be that way, and evolution is possible because the properties of DNA and related molecules allow organisms to pass on their genetic blueprint to their offspring. In precisely the same sense, whether or not the *discoveries* of elementary particle physics are useful to all other scientists, the *principles* of elementary particle physics are fundamental to all nature.

Opponents of reductionism often rely on the argument that discoveries in elementary particle physics are not likely to be useful to scientists in other fields. This is not supported by the evidence of history. The elementary particle physics of the first half of this century was largely the physics of electrons and photons, which had an enormous and unquestioned effect on our understanding of matter in all its forms. The discoveries of today's elementary particle physics are already having an important influence on cosmology and astronomy—for instance, we use our knowledge of the menu of elementary particles in calculating the production of chemical elements in the first few minutes of the universe. No one can say what other consequences these discoveries may have.

But, just for argument, suppose for a moment that *no* discovery made by elementary particle physicists would ever again affect the work of scientists in any other fields. The work of elementary particle physics would still have a special importance. We know that the evolution of living things has been made possible by the properties of DNA and other molecules and that the properties of any molecule are what they are because of the properties of electrons and atomic nuclei and electric forces. And why are these things the way they are? This has been partly explained by the standard model of elementary particles, and now we want to take the next step and explain the standard model and the principles of relativity and other symmetries on which it is based. I do not understand how this cannot seem an important task to anyone who is curious about why the world is the way it is, quite apart from any possible use that elementary particle physics may have to any other scientists.

Indeed, elementary particles are not in themselves very interesting, not at any rate in the way that people are interesting. Aside from their momentum and spin, every electron in the universe is just like every other electron—if you have seen one electron, you have seen them all. But this very simplicity suggests that electrons, unlike people, are not made up of numbers of more fundamental constituents, but are themselves something close to the fundamental constituents of everything else. It is because elementary particles are so boring that they are interesting; their simplicity suggests that the study of elementary particles will bring us closer to a comprehensive understanding of nature.

The example of high-temperature superconductivity may help to explain the special and limited sense in which elementary particle physics is more fundamental than other branches of physics. Right now, Anderson and other condensed matter physicists are trying to understand the puzzling persistence of superconductivity in certain compounds of copper, oxygen, and more exotic elements at temperatures far above what had been thought possible. At the same time, elementary particle physicists are trying to understand the origin of the masses of quarks and electrons and other particles in the standard model. (The two problems happen to be related mathematically; as we shall see, they both boil down to a question of how certain symmetries of the underlying equations become lost in the solutions of the equations.) Condensed matter physicists will doubtless eventually solve the problem of high-temperature superconductivity without any direct help from elementary particle physicists, and, when elementary particle physicists understand the origin of mass, it will very likely be without direct inputs from condensed matter physics. The difference between these two problems is that, when condensed matter physicists finally explain high-temperature superconductivity—whatever brilliant new ideas have to be invented along the way—in the end the explanation will take the form of a mathematical demonstration that deduces the existence of this phenomenon from *known* properties of electrons, photons, and atomic nuclei; in contrast, when particle physicists finally understand the origin of mass in the standard model the explanation will be based on aspects of standard model about which we are today quite uncertain, and which we cannot learn (though we may guess) without new data from facilities like the Super Collider. Elementary particle physics thus represents a frontier of our knowledge in a way that condensed matter physics does not.

This does not in itself solve the problem of how to allocate research funds. There are many motives for doing scientific research—applications to medicine and technology, national prestige, mathematical athleticism, and the sheer joy of understanding beautiful phenomena—that are fulfilled by other sciences as well as (and sometimes better than) by elementary particle physics. Particle physicists do not believe that the uniquely fundamental character of their work gives them a first charge on the public purse, but they do believe that it is not a factor that can be simply ignored in decisions about the support of scientific research.

Possibly the best-known attempt to set standards for making this sort of decision is that of Alvin Weinberg.* In a 1964 article he offered this guideline: "I would therefore sharpen the criterion of scientific merit by proposing that, other things being equal, *that field has the most scientific merit which contributes most heavily to and illuminates most brightly its neighboring scientific disciplines*" (his italics). After reading an article of mine on these issues, Alvin wrote to me to remind me of his proposal. I had not forgotten it, but I had not agreed with it either. As I wrote in reply to Alvin, this sort of reasoning could be used to justify spending billions of dollars in classifying the butterflies of Texas, on the grounds that this would illuminate the classification of the butterflies of Oklahoma, and indeed of butterflies in general. This silly example was meant only to illustrate that it does not add much to the importance of an uninteresting scientific project to say that it is important to other uninteresting scientific projects. (I am probably now in trouble with lepidopterists who would like to spend billions of dollars classifying the butterflies of Texas.) But what I really miss in Alvin Weinberg's criteria for scientific choice is the lack of the *reductionist* perspective; that one of the things that makes some work in science interesting is that it takes us closer to the point where all our explanations converge.

Some of the issues in the debate over reductionism within physics have been usefully raised by the author James Gleick. (It was Gleick who introduced the physics of chaos to a general readership.) In a recent talk he argued:

---

* Alvin Weinberg and I are friends, but we are not related. In 1966 when I was first visiting Harvard I found myself at lunch at the faculty club with the late John Van Vleck, a crusty and patrician older physicist who had been one of the first to apply the new methods of quantum mechanics to the theory of the solid state in the late 1920s. Van Vleck asked me if I was related to "the" Weinberg. I was a bit put out, but I understood what he meant; I was at that time a rather junior theorist, and Alvin was director of the Oak Ridge National Laboratory. I dipped into my reserves of effrontery, and replied that I was "the" Weinberg. I do not think that Van Vleck was impressed.

Chaos is anti-reductionist. This new science makes a strong claim about the world: namely, that when it comes to the most interesting questions, questions about order and disorder, decay and creativity, pattern formation and life itself, the whole cannot be explained in terms of the parts.

There are fundamental laws about complex systems, but they are new kinds of laws. They are laws of structure and organization and scale, and they simply vanish when you focus on the individual constituents of a complex system—just as the psychology of a lynch mob vanishes when you interview individual participants.

I would reply first that different questions are interesting in different ways. Certainly questions about creativity and life are interesting because we are alive and would like to be creative. But there are other questions that are interesting because they carry us closer to the point of convergence of our explanations. Discovering the source of the Nile did nothing to illuminate the problems of Egyptian agriculture, but who can say that it was not interesting?

Also, it misses the point of this sort of question to speak of explaining the whole "in terms of the parts"; the study of quarks and electrons is fundamental not because all ordinary matter is composed of quarks and electrons but because we think that by studying quarks and electrons we will learn something about the *principles* that govern everything. (It was an experiment using electrons fired at the quarks inside atomic nuclei that clinched the case for the modern unified theory of two of the four fundamental forces of nature, the weak and electromagnetic forces.) In fact the elementary particle physicist today gives more attention to exotic particles that are *not* present in ordinary matter than to the quarks and electrons that are, because we think that right now the questions that need to be answered will be better illuminated by studying these exotic particles. When Einstein explained the nature of gravitation in his general theory of relativity, it was not "in terms of the parts," but in terms of the geometry of space and time. It may be that physicists in the twenty-first century will find that the study of black holes or gravitational radiation reveals more about the laws of nature than elementary particle physics does. Our present concentration on elementary particles is based on a tactical judgment—that at *this* moment in the history of science this is the way to make progress toward the final theory.

Finally, there is the question of emergence: is it really true that there are new kinds of laws that govern complex systems? Yes, of course, in the sense that different levels of experience call for description and analysis in different terms. The same is just as true for chemistry as for chaos. But *fundamental* new kinds of laws? Gleick's lynch mob provides a counterexample. We may formulate what we learn about mobs in the form of laws (such as the old saw that revolutions always eat their children), but, if we ask for an explanation of why such laws hold, we would not be very happy to be told that the laws are fundamental, without explanation in terms of anything else. Rather, we would seek a reductionist explanation precisely in terms of the psychology of individual humans. The same is true for the emergence of chaos. The exciting progress that has been made in this area in recent years has not taken the form solely of the observation of chaotic systems and the formulation of empirical laws that describe them; even more important has been the mathematical deduction of the laws governing chaos from the microscopic physical laws governing the systems that become chaotic.

I suspect that all working scientists (and perhaps most people in general) are in practice just as reductionist as I am, although some like Ernst Mayr and Philip Anderson do not like to express themselves in these terms. For instance, medical research deals with problems that are so urgent and difficult that proposals of new cures often must be based on medical statistics, without understanding how the cure works, but even if a new cure were suggested by experience with many patients, it would probably be met with skepticism if one could not see how it could possibly be explained reductively, in terms of sciences like biochemistry and cell biology. Suppose that a medical journal carried two articles reporting two different cures for scrofula: one by ingestion of chicken soup and the other by a king's touch. Even if the statistical evidence presented for these two cures had equal weight, I think that the medical community (and everyone else) would have very different reactions to the two articles. Regarding chicken soup I think that most people would keep an open mind, reserving judgment until the cure could be confirmed by independent tests. Chicken soup is a complicated mixture of good things, and who knows what effect its contents might have on the mycobacteria that cause scrofula? On the other hand whatever statistical evidence were offered to show that a king's touch helps to cure scrofula, readers would tend to be very skeptical, suspecting a hoax or a meaningless coincidence, because they would see no way that such a cure could ever be explained reductively. How could it matter to a mycobacterium whether the person touching its host was properly crowned and anointed or the eldest son of the previous monarch? (Even in the Middle Ages when it was commonly believed that a king's touch would cure scrofula, kings themselves seemed to have had their doubts about this. As far as I know, in all the medieval struggles over disputed successions, as between Plantagenet and Valois or York and Lancaster, no claimant to a throne ever tried to prove his title by demonstrating the curative power of his touch.) A biologist today who protested that this cure needed no explanation because the power of the king's touch is an autonomous law of nature, as fundamental as any other, would not get much encouragement from his colleagues, because they are guided by a reductionist worldview that has no place for such an autonomous law.

The same is true throughout all the sciences. We would not pay much attention to a proposed autonomous law of macroeconomics that could not possibly be explained in terms of the behavior of individuals or to a hypothesis about superconductivity that could not possibly be explained in terms of the properties of electrons and photons and nuclei. The reductionist attitude provides a useful filter that saves scientists in all fields from wasting their time on ideas that are not worth pursuing. In this sense, we are all reductionists now.

CHAPTER IV

QUANTUM MECHANICS AND ITS DISCONTENTS

A player put a ball on the table and hit it with the cue. Watching the rolling ball, Mr. Tompkins noticed to his great surprise that the ball began to "spread out." This was the only expression he could find for the strange behavior of the ball which, moving across the green field, seemed to become more and more washed out, losing its sharp contours. It looked as if not one ball was rolling across the table but a great many balls, all partially penetrating into each other. Mr. Tompkins had often observed analogous phenomena before, but today he had not taken a single drop of whisky and he could not understand why it was happening now.

George Gamow, *Mr. Tompkins in Wonderland*

The discovery of quantum mechanics in the mid-1920s was the most profound revolution in physical theory since the birth of modern physics in the seventeenth century. In the account earlier of the properties of a piece of chalk, our chains of questions led us again and again to answers in terms of quantum mechanics. All the fancy mathematical theories that physicists have pursued in recent years—quantum field theories, gauge theories, superstring theories—are formulated within the framework of quantum mechanics. If there is anything in our present understanding of nature likely to survive in a final theory, it is quantum mechanics.

The historical importance of quantum mechanics lies not so much in the fact that it provided answers to a number of old questions about the nature of matter—much more important is that it changed our idea of the questions that we are allowed to ask. For Newton's physicist successors, physical theories were intended to provide a mathematical machine that would allow physicists to calculate the positions and velocities of the particles of any system at all future times from a complete knowledge (never of course realized in practice) of their values at any one instant. But quantum mechanics introduced a completely new way of talking about the state of a system. In quantum mechanics we speak of mathematical constructs called wave functions that give us information only about the probabilities of various possible positions and velocities. So profound is this change, that physicists now use the word "classical" to mean not "Greco-Roman," or "Mozart, etc.," but, rather, "before quantum mechanics."

If there is any moment that marks the birth of quantum mechanics, it would be a vacation taken by the young Werner Heisenberg in 1925. Suffering from hay fever, Heisenberg fled the flowering fields near Göttingen for the lonely North Sea island of Helgoland. Heisenberg and his colleagues had for several years been struggling with a problem raised in 1913 by Niels Bohr's theory of the atom: why do electrons in atoms occupy only certain allowed orbits with certain definite energies? On Helgoland Heisenberg made a fresh start. He decided that, because no one could ever directly observe the orbit of an electron in an atom, he would deal only with quantities that could be measured: specifically, with the energies of the quantum *states* in which all the atom's electrons occupy allowed orbits, and with the rates at which an atom might spontaneously make a transition from any one of these quantum states to any other state by emitting a particle of light, a photon. Heisenberg made what he called a "table" from these rates, and he introduced mathematical operations on this table that would yield new tables, one type of table for each physical quantity like the position or the velocity or the square of the velocity of an electron.* Knowing how the energy of a particle in a simple system depends on its velocity and position, Heisenberg was able in this way to calculate a table of the energies of the system in its various quantum states, in a sort of parody of the way that the energy of a planet is calculated in Newton's physics from a knowledge of its position and velocity.

If the reader is mystified at what Heisenberg was doing, he or she is not alone. I have tried several times to read the paper that Heisenberg wrote on returning from Helgoland, and, although I think I understand quantum mechanics, I have never understood Heisenberg's motivations for the mathematical steps in his paper. Theoretical physicists in their most successful work tend to play one of two roles: they are either *sages* or *magicians*. The sage-physicist reasons in an orderly way about physical problems on the basis of fundamental ideas of the way that nature ought to be. Einstein, for example, in developing the general theory of relativity, was playing the role of a sage; he had a well-defined problem—how to fit the theory of gravita-

---

* More accurately, the entries in Heisenberg's table were what are called transition amplitudes, quantities whose squares give the transition rates. Heisenberg was told after he returned to Göttingen from Heligoland that his mathematical operations on these tables were already well known to mathematicians; such tables were known to mathematicians as matrices and the operation by which one goes from the table representing the velocity of an electron to the table representing its square was known as matrix multiplication. This is one example of the spooky ability of mathematicians to anticipate structures that are relevant to the real world.

tion into the new view of space and time that he had proposed in 1905 as the special theory of relativity. He had some valuable clues, in particular the remarkable fact discovered by Galileo that the motion of small bodies in a gravitational field are independent of the nature of the bodies. This suggested to Einstein that gravitation might be a property of space-time itself. Einstein also had available a well-developed mathematical theory of curved spaces that had been worked out by Riemann and other mathematicians in the nineteenth century. It is possible to teach general relativity today by following pretty much the same line of reasoning that Einstein used when he finally wrote up his work in 1915. Then there are the magician-physicists, who do not seem to be reasoning at all but who jump over all intermediate steps to a new insight about nature. The authors of physics textbooks are usually compelled to redo the work of the magicians so that they seem like sages; otherwise no reader would understand the physics. Planck was a magician in inventing his 1900 theory of heat radiation, and Einstein was playing the part of a magician when he proposed the idea of the photon in 1905. (Perhaps this is why he later described the photon theory as the most revolutionary thing he had ever done.) It is usually not difficult to understand the papers of sage-physicists, but the papers of magician-physicists are often incomprehensible. In this sense, Heisenberg's 1925 paper was pure magic.

Perhaps we should not look too closely at Heisenberg's first paper. Heisenberg was in touch with a number of gifted theoretical physicists, including Max Born and Pascual Jordan in Germany and Paul Dirac in England, and before the end of 1925 they had fashioned Heisenberg's ideas into an understandable and systematic version of quantum mechanics, today called matrix mechanics. By the following January Heisenberg's old classmate Wolfgang Pauli in Hamburg was able to use the new matrix mechanics to solve the paradigmatic problem of atomic physics, the calculation of the energies of the quantum states of the hydrogen atom, and thus to justify Bohr's earlier ad hoc results.

The quantum-mechanical calculation of the hydrogen energy levels by Pauli was an exhibition of mathematical brilliance, a sagelike use of Heisenberg's rules and the special symmetries of the hydrogen atom. Although Heisenberg and Dirac may have been even more creative than Pauli, no physicist alive was more clever. But not even Pauli was able to extend his calculation to the next simplest atom, that of helium, much less to heavier atoms or molecules.

The quantum mechanics taught in undergraduate courses and used in their everyday work by chemists and physicists today is not in fact the matrix mechanics of Heisenberg and Pauli and their collaborators, but a mathematically equivalent— though far more convenient—formalism introduced a little later by Erwin Schrödinger. In Schrödinger's version of quantum mechanics each possible physical state of a system is described by giving a quantity known as the *wave function* of the system, in something like the way that light is described as a wave of electric and magnetic fields. The wave-function approach to quantum mechanics had appeared before the work of Heisenberg, in 1923 papers of Louis de Broglie, and in his 1924 Paris doctoral thesis. De Broglie guessed that the electron can be regarded as some sort of wave, with a wavelength that is related to the electron's momentum in the same way that light wavelengths according to Einstein are related to the photons' momentum: the wavelength in both cases is equal to a fundamental constant of nature known as Planck's constant, divided by the momentum. De Broglie did not have any idea of the physical significance of the wave, and he did not invent any sort of dynamical wave equation; he simply assumed that the allowed orbits of the electrons in a hydrogen atom would have to be just large enough for some number of complete wavelengths to fit around the orbit: one wavelength for the lowest energy state, two wavelengths for the next lowest, and so on. Remarkably, this simple and not very well motivated guess gave the same successful answers for the energies of the orbits of the electron in the hydrogen atom as Bohr's calculation a decade earlier.

With such a doctoral thesis behind him, it might have been expected that de Broglie would go on to solve all the problems of physics. In fact he did virtually nothing else of scientific importance throughout his life. It was Schrödinger in Zurich who in 1925–26 transformed de Broglie's rather vague ideas about electron waves into a precise and coherent mathematical formalism that applied to electrons or other particles in any sort of atom or molecule. Schrödinger was also able to show that his "wave mechanics" is equivalent to Heisenberg's matrix mechanics; either can be deduced mathematically from the other.

At the heart of Schrödinger's approach was a dynamical equation (known ever since as the Schrödinger equation) that dictated the way that any given particle wave would change with time. Some of the solutions of the Schrödinger equation for electrons in atoms simply oscillate at a single pure frequency, like the sound wave produced by a perfect tuning fork. Such special solutions correspond to the possible stable quantum states of the atom or molecule (something like the stable waves of vibration within a tuning fork), with the energy of the atomic state given by the frequency of the wave times Planck's constant. These are the energies that are revealed to us through the colors of the light that the atom can emit or absorb.

The Schrödinger equation is mathematically the same sort of equation (known as a partial differential equation) that had been used since the nineteenth century to study waves of sound or light. Physicists in the 1920s were already so comfortable with this sort of wave equation that they were able immediately to set about calculating the energies and other properties of all sorts of atoms and molecules. It was a golden time for physics. Other successes followed rapidly, and one by one the mysteries that had surrounded atoms and molecules seemed to melt away.

Despite this success, neither de Broglie nor Schrödinger nor anyone else at first knew what sort of physical quantity was oscillating in an electron wave. Any sort of wave is described at any one moment by a list of numbers, one number for each point of the space through which the wave passes. For instance, in a sound wave the numbers give the air pressure at each point of the air. In a light wave, the numbers give the strengths and directions of the electric and magnetic fields at each point of the space through which the light travels. The electron wave could also be described at any moment as a list of numbers, one number for each point of the space in and around the atom. It is this list that is known as the wave function, and the individual numbers are called the values of the wave function. But at first all one could say about the wave function is that it was a solution of the Schrödinger equation; no one yet knew what physical quantity was being described by these numbers.

The quantum theorists of the mid-1920s were in the same position as the physicists studying light at the beginning of the nineteenth century. The observation of phenomena like diffraction (the failure of light rays to follow straight lines when passing very close to objects or through very small holes) had suggested to Thomas Young and Augustin Fresnel that light was some sort of wave and that it did not travel in straight lines when forced to squeeze through small holes because the holes were smaller than its wavelength. But no one in the early nineteenth century knew what light was a wave *of;* only with the work of James Clerk Maxwell in the 1860s did it become clear that light was a wave of varying electric and magnetic fields. But what is it that is varying in an electron wave?

The answer came from a theoretical study of how free electrons behave when they are fired at atoms. It is natural to describe an electron traveling through empty space as a wave packet, a little bundle of electron waves that travel along together, like the pulse of light waves produced by a searchlight that is turned on only for an instant. The Schrödinger equation shows that, when such a wave packet strikes an atom, it breaks up; wavelets go traveling off in all directions like sprays of water when the stream from a garden hose hits a rock. This was puzzling; electrons striking atoms fly off in one direction or another but they do not break up—they remain electrons. In 1926 Max Born in Göttingen proposed to interpret this peculiar behavior of the wave function in terms of probabilities. The electron does not break up, but it can be scattered in any direction, and the probability that an electron is scattered in some particular direction is greatest in those directions where the values of the wave function are largest. In other words, electron waves are not waves *of* anything; their significance is simply that the value of the wave function at any point tells us the probability that the electron is at or near that point.

Neither Schrödinger nor de Broglie were comfortable with this interpretation of electron waves, which probably explains why neither of them contributed importantly to the further development of quantum mechanics. But the probabilistic interpretation of the electron waves found support in a remarkable argument offered by Heisenberg the following year. Heisenberg considered the problems that are encountered when a physicist sets out to measure the position and momentum of an electron. In order to make an accurate measurement of position it is necessary to use light of short wavelength, because diffraction always blurs images of anything smaller than a wavelength of light. But light of short wavelength consists of photons with correspondingly high momentum, and, when photons of high momentum are used to observe an electron, the electron necessarily recoils from the impact, carrying off some fraction of the photon's momentum. Thus the more accurately we try to measure the position of an electron, the less we know after the measurement about the electron's momentum. This rule has come to be known as the *Heisenberg uncertainty principle.** An electron wave that is very sharply peaked at some position represents an electron that has a fairly definite position but a momentum that could have almost any value. In contrast, an electron wave that takes the form of a smooth, equally spaced alternation of crests and troughs extending over many wavelengths represents an electron that has a fairly definite momentum but whose position is highly uncertain. More typical electrons like those in atoms or molecules have neither a definite position nor momentum.

Physicists continued to wrangle over the interpretation of quantum mechanics for years after they had become used to solving the Schrödinger equation. Einstein was unusual in rejecting quantum mechanics in his work; most physicists were simply trying to understand it. Much of this debate went on at the University Institute for Theoretical Physics in Copenhagen,

---

* To be a little more precise, because the wavelength of light equals Planck's constant divided by the photon momentum, the uncertainty in any particle's position cannot be less than Planck's constant divided by the uncertainty in its momentum. We do not notice this uncertainty for ordinary objects like billiard balls because Planck's constant is so small. In the system of units with which physicists are most familiar, based on the centimeter, gram, and second as the fundamental units of length, mass, and time, Planck's constant is 6.626 thousandth millionth millionth millionth millionths, a decimal point followed by twenty-six zeros and then 6626. Planck's constant is so small that the wavelength of a billiard ball rolling across a table is much less than the size of an atomic nucleus, so there is no difficulty in making quite accurate measurements of the ball's position and momentum at the same time.

under the guidance of Niels Bohr.* Bohr focused particularly on a peculiar feature of quantum mechanics that he called *complementarity:* knowledge of one aspect of a system precludes knowledge of certain other aspects of the system. Heisenberg's uncertainty principle provides one example of complementarity: knowledge of a particle's position (or momentum) precludes knowledge of the particle's momentum (or position).†

By around 1930 the discussions at Bohr's institute had led to an orthodox "Copenhagen" formulation of quantum mechanics, in terms that were now much more general than the wave mechanics of single electrons. Whether a system consists of one or many particles, its state at any moment is described by the list of numbers known as the values of the wave function, one number corresponding to each possible configuration of the system. The same state may be described by giving the values of the wave function for configurations that are characterized in various different ways—for instance, by the positions of all the particles of the system, or by the momenta of all the particles of the system, or in various other ways, though not by the positions *and* the momenta of all the particles.

The essence of the Copenhagen interpretation is a sharp separation between the system itself and the apparatus used to

---

*I had the good fortune to meet Bohr, though near the end of his career and the beginning of mine. Bohr was my host when I went for my first year of graduate studies to his institute in Copenhagen. However, we spoke only briefly, and I took away no words of wisdom—Bohr was a famous mumbler, and it was always hard to tell what he meant. I remember the horrified look on my wife's face when Bohr spoke to her at length at a party in the conservatory of his house, and she realized she was missing everything that the great man was saying.
†In later years Bohr emphasized the importance of complementarity for matters far removed from physics. There is a story that Bohr was once asked in German what is the quality that is complementary to truth (*wahrheit*). After some thought he answered clarity (*klarheit*). I have felt the force of this remark in writing the present chapter.

measure its configuration. As Max Born had emphasized, during the times between measurements the values of the wave function evolve in a perfectly continuous and deterministic way, dictated by some generalized version of the Schrödinger equation. While this is going on, the system cannot be said to be in any definite configuration. If we measure the configuration of the system (e.g., by measuring all the particles' positions *or* all their momenta, but not both), the system jumps into a state that is definitely in one configuration or another, with probabilities given by the squares of the values of the wave function for these configurations just before the measurement.

Describing quantum mechanics in words alone inevitably gives only a vague impression of what it is about. Quantum mechanics itself is not vague; although it seems weird at first, it provides a precise framework for calculating energies and transition rates and probabilities. I want to try to take the reader a little further into quantum mechanics, and for this purpose I shall consider here the simplest possible sort of system, one that has just two possible configurations. We can think of this system as a mythical particle with only two instead of an infinite number of possible positions—say, *here* and *there*. The state of the system at any moment is then described by two numbers: the *here* and *there* values of the wave function.

In classical physics the description of our mythical particle is very simple: it is definitely either *here* or *there*, though it may jump from *here* to *there* or vice versa in a manner dictated by some dynamical law. Matters are more complicated in quantum mechanics. When we are not observing the particle, the state of the system could be pure *here*, in which case the *there* value of the wave function would vanish, or pure *there*, in which case the *here* value of the wave function would vanish, but it is also possible (and more usual) that neither value vanishes and that the particle is neither definitely *here* nor definitely *there*. If we do look to see whether the particle is *here* or *there,* we of course find that it is at one or the other position; the probability that it will turn out to be *here* is given by the square of the *here* value just before the measurement, and the probability that it is *there* is given by the square of the *there* value. According to the Copenhagen interpretation, when we measure whether the particle is in the *here* or *there* configuration, the values of the wave function jump to new values; either the *here* value becomes equal to one and the *there* value becomes equal to zero, or vice versa, but knowing the wave function we cannot predict which will happen, only their probabilities.

This system with only two configurations is so simple that its Schrödinger equation can be described without symbols. In between measurements the rate of change of the *here* value of the wave function is a certain constant number times the *here* value plus a second constant number times the *there* value; the rate of change of the *there* value is a third constant number times the *here* value plus a fourth constant number times the *there* value. These four constant numbers are collectively known as the *Hamiltonian* of this simple system. The Hamiltonian characterizes the system itself rather than any particular state of the system; it tells us everything there is to know about how the state of the system evolves from any given initial conditions. By itself quantum mechanics does not tell us what the Hamiltonian is—the Hamiltonian has to be derived from our experimental and theoretical knowledge about the nature of the system in question.

This simple system can incidentally also be used to illustrate Bohr's idea of complementarity, by considering other ways of describing the state of the same particle. For instance, there is a pair of states, something like states of definite momentum, that we may call *stop* and *go,* in which the *here* value of the wave function is respectively either equal to the *there* value, or equal to minus the *there* value. We can if we like describe the wave function in terms of its *stop* and *go* values rather than its *here* and *there* values: the *stop* value is the sum of the *here* and *there* values, and the *go* value is the difference. If we happen to know that the particle's position is definitely *here* then the *there* value of the wave function must vanish and so the *stop* and *go* values of the wave function must be equal, which means that we know nothing about the particle's momentum; both possibilities have 50% probability. Conversely, if we know that the particle is definitely in the state *stop* with zero momentum then the *go* value of the wave function vanishes, and, because the *go* value is the difference of the *here* and *there* values, the *here* and *there* values must then be equal, which means that we know nothing about whether the particle is *here* or *there;* the probability of either is 50%. We see that there is a complete complementarity between a *here*-or-*there* and a *stop*-or-*go* measurement: we can make either sort of measurement, but whichever one we choose to make leaves us completely in the dark about the results we would find if we made the other sort of measurement.

Everyone agrees on how to use quantum mechanics, but there is serious disagreement about how to think about what we are doing when we use it. For some who felt wounded by the reductionism and determinism of Newtonian physics, two aspects of quantum mechanics seemed to offer a welcome balm. Where human beings had no special status in Newtonian physics, in the Copenhagen interpretation of quantum mechanics humans play an essential role in giving meaning to the wave function by the act of measurement. And where the Newtonian physicist spoke of precise predictions the quantum mechanician now offers only calculations of probabilities, thus seeming to make room again for human free will or divine intervention.

Some scientists and writers like Fritjof Capra welcome what they see as an opportunity for a reconciliation between the spirit of science and the gentler parts of our nature. I might, too, if I thought the opportunity was a real one, but I do not think it is. Quantum mechanics has been overwhelmingly important to physics, but I cannot find any messages for human life in quantum mechanics that are different in any important way from those of Newtonian physics.

Because these matters are still controversial, I have persuaded two well-known figures to debate them here.

A DIALOGUE ON THE MEANING OF
QUANTUM MECHANICS

TINY TIM: I think quantum mechanics is just wonderful. I never did like the way that in Newtonian mechanics if you knew the position and velocity of every particle at one moment you could predict everything about the future, with no room for free will and no special role for humans at all. Now in quantum mechanics all your predictions are vague and probabilistic, and nothing has a definite state until human beings observe it. I'm sure that some Eastern mystic must have said something like this.

SCROOGE: Bah! I may have changed my mind about Christmas, but I still know humbug when I hear it. It is true enough that the electron does not have a definite position and momentum at the same time, but this just means that these are not the appropriate quantities to use in describing the electron. What an electron or any collection of particles does have at any time is a wave function. If there is a human observing the particles, then the state of the whole system including the human is described by a wave function. The evolution of the wave function is just as deterministic as the orbits of particles in Newtonian mechanics. In fact, it is more deterministic, because the equations that tell us how the wave function develops over time are too simple to allow chaotic solutions. Where's your free will now?

TINY TIM: I am really surprised that you should reply in such an unscientific way. The wave function has no objective reality, because it cannot be measured. For instance, if we observe that a particle is *here,* we cannot conclude from this that the wave function *before* the observation had a vanishing *there* value; it might have had any *here* and *there* values, with the particle just happening to show up as *here* rather than *there* when it was observed. If the wave function is not real, then why are you making so much of the fact that it evolves deterministically? All that we ever measure are quantities like positions or momenta or spins, and about these we can predict only probabilities. And until some human intervenes to measure these quantities, we cannot say that the particle has any definite state at all.

SCROOGE: My dear young person, you seem to have swallowed uncritically the nineteenth-century doctrine called positivism, which says that science should concern itself only with things that can actually be observed. I agree that it is not possible to measure a wave function in any one experiment. So what? By repeating measurements many times for the same initial state, you can work out what the wave function in the state must be and use the results to check our theories. What more do you want? You really should bring your thinking up to the twentieth century. Wave functions are real for the same reason that quarks and symmetries are—because it is useful to include them in our theories. Any system is in a definite state *whether any humans are observing it or not;* the state is not described by a position or a momentum but by a wave function.

TINY TIM: I do not think I want to argue about what is real or not with someone who spends his evenings wandering about with ghosts. Let me just remind you of a serious problem you get into when you imagine the wave function to be real. This problem was mentioned in an attack on quantum mechanics by Einstein at the 1933 Solvay Conference in Brussels and then in 1935 written up by him in a famous paper with Boris Podolsky and Nathan Rosen. Suppose that we have a system consisting of two electrons, prepared in such a way that the electrons at some moment have a known large separation and a known total momentum. (This does not violate the Heisenberg uncertainty principle. For instance, the separation could be measured as accurately as you like by sending light rays of very short wavelength from one electron to the other; this would disturb the momentum of each electron but because of the conservation of momentum it would not change their *total* momentum.) If someone then measures the momentum of the first electron the momentum of the second electron could be calculated immediately because their sum is known. On the other hand, if someone measures the position of the first electron then the position of the second electron could be calculated immediately because their separation is known. But this means that by observing the state of the first electron you can instantaneously change the wave function so that the second electron has a definite position or a definite momentum, *even though you never come anywhere near the second electron*. Are you really happy thinking of wave functions as real if they can change in this way?

SCROOGE: I can accept it. Nor do I worry about the rule of special relativity that forbids sending signals faster than the speed of light; there is no conflict with this rule. A physicist who measures the momentum of the second electron has no way of knowing that the value she finds has been affected by observations of the first electron. For all she knows, the electron before her measurement might just as well have had a definite position as a definite momentum. Not even Einstein could use this sort of measurement to send instantaneous signals from one electron to the other. (While you were at it, you might have mentioned that John Bell has come up with even weirder consequences of quantum mechanics involving atomic spins, and experimental physicists have demonstrated that the spins in atomic systems really do behave in the way expected from quantum mechanics, but that is just the way the world is.) It seems to me that none of this forces us to stop thinking of the wave function as real; it just behaves in ways that we are not used to, including instantaneous changes affecting the wave function of the whole universe. I think that you should stop looking for profound philosophical messages in quantum mechanics, and let me get on with using it.

TINY TIM: With the greatest respect, I must say that if you can accept instantaneous changes in the wave function throughout all space, I suspect you can accept anything. Anyway, I hope you will forgive me when I say that you are not being very consistent. You have said that the wave function of any system evolves in a perfectly deterministic way, and that probabilities come into the picture only when we make measurements. But according to your point of view not only the electron but also the measuring apparatus and the human observer who uses it are all just one big system, described by a wave function with a vast number of values, all of which evolve deterministically even during a measurement. So if everything happens deterministically, how can there be any uncertainty about the results of measurements? Where do the probabilities come from when measurements are made?

———

I have some sympathy with both sides in this debate, though rather more with the realist Scrooge than with the positivist Tiny Tim. I gave Tiny Tim the last word because the problem

he finally raises has been the most important puzzle in the interpretation of quantum mechanics. The orthodox Copenhagen
interpretation that I have been describing up to now is based on
a sharp separation between the physical system, governed by
the rules of quantum mechanics, and the apparatus used to
study it, which is described classically, that is, according to the
prequantum rules of physics. Our mythical particle may have a
wave function with both *here* and *there* values, but, when it is
observed, it somehow becomes definitely *here* or *there,* in a
manner that is essentially unpredictable, except with regard to
probabilities. But this difference of treatment between the system being observed and the measuring apparatus is surely a fiction. We believe that quantum mechanics governs everything in
the universe, not only individual electrons and atoms and molecules but also the experimental apparatus and the physicists
who use it. If the wave function describes the measuring apparatus as well as the system being observed and evolves *deterministically* according to the rules of quantum mechanics even
during a measurement, then, as Tiny Tim asks, where do the
probabilities come from?

Dissatisfaction with the artificial split between systems and
observers in the Copenhagen interpretation has led many theorists to a rather different point of view, the *many-worlds* or
many-histories interpretation of quantum mechanics, first presented in the Princeton Ph.D. thesis of Hugh Everett. According
to this view, a *here*-or-*there* measurement of our mythical particle consists of some sort of interaction between the particle
and the measuring apparatus, such that the wave function of
the combined system winds up with appreciable values for just
two configurations; one value corresponds to the configuration
in which the particle is *here* and the dial on the apparatus points
to *here;* the other corresponds to the possibility that the particle
is *there* and the dial on the apparatus points to *there.* There is

still a definite wave function, produced in a completely deterministic way by the interaction of the particle with the measuring apparatus as governed by the rules of quantum mechanics. However, the two values of the wave function correspond to states of different energy, and, because the measuring apparatus is macroscopic, this energy difference is very large, so that these two values oscillate at very different frequencies. Observing the position of the dial on the measuring apparatus is like tuning in at random to either one or the other of two radio stations, KHERE and WTHERE; as long as the broadcast frequencies are well separated, there is no interference, and you receive one station or the other, with probabilities proportional to their intensities. The absence of interference between the two values of the wave function means that in effect the world's history has been split into two separate histories, in one of which the particle is *here* and in the other *there,* and each of these two histories will thenceforth unfold without interaction with the other.

By applying the rules of quantum mechanics to the combined system of particle and measuring apparatus, one can actually prove that the probability of finding the particle *here* with the dial on the apparatus pointing to *here* is proportional to the square of the *here* value of the wave function of the particle just before it began to interact with the measuring apparatus, just as postulated in the Copenhagen interpretation of quantum mechanics. However, this argument does not really answer Tiny Tim's question. In calculating the probability of the combined system of particle and measuring apparatus being in any one configuration, we are implicitly dragging in an observer who reads the dial and finds that it reads *here* or *there*. Although in this analysis the measuring apparatus was treated quantum mechanically, the observer was treated classically; she finds that the dial definitely points either to *here* or *there,* in a manner that again cannot be predicted except as to probabilities. We could of course treat the observer quantum mechanically, but only at the cost of introducing another observer who detects the conclusions of the first one, perhaps by reading an article in a physics journal. And so on.

A long line of physicists have worked to purge the foundations of quantum mechanics of any statement about probabilities or any other interpretive postulate that draws a distinction between systems and observers. What one needs is a quantum-mechanical model with a wave function that describes not only various systems under study but also something representing a conscious observer. With such a model, one would try to show that, as a result of *repeated* interactions of the observer with individual systems, the wave function of the combined system evolves with certainty to a final wave function, in which the observer has become convinced that the probabilities of the individual measurements are what are prescribed in the Copenhagen interpretation. I am not convinced that this program has been entirely successful yet, but I think in the end it may be. If so, then Scrooge's realism will be entirely vindicated.

It is truly surprising how little difference all this makes. Most physicists use quantum mechanics every day in their working lives without needing to worry about the fundamental problem of its interpretation. Being sensible people with very little time to follow up all the ideas and data in their own specialties and not having to worry about this fundamental problem, they do not worry about it. A year or so ago, while Philip Candelas (of the physics department at Texas) and I were waiting for an elevator, our conversation turned to a young theorist who had been quite promising as a graduate student and who had then dropped out of sight. I asked Phil what had interfered with the ex-student's research. Phil shook his head sadly and said, "He tried to understand quantum mechanics."

So irrelevant is the philosophy of quantum mechanics to its

use, that one begins to suspect that all the deep questions about the meaning of measurement are really empty, forced on us by our language, a language that evolved in a world governed very nearly by classical physics. But I admit to some discomfort in working all my life in a theoretical framework that no one fully understands. And we really do need to understand quantum mechanics better in quantum cosmology, the application of quantum mechanics to the whole universe, where no outside observer is even imaginable. The universe is much too large now for quantum mechanics to make much difference, but according to the big-bang theory there was a time in the past when the particles were so close together that quantum effects must have been important. No one today knows even the rules for applying quantum mechanics in this context.

Of even greater interest it seems to me is the question of whether quantum mechanics is necessarily *true*. Quantum mechanics has had phenomenal successes in explaining the properties of particles and atoms and molecules, so we know that it is a very good approximation to the truth. The question then is whether there is some other logically possible theory whose predictions are very close but not quite the same as those of quantum mechanics. It is easy to think of ways of changing most physical theories in small ways. For instance, the Newtonian law of gravitation, that the gravitational force between two particles decreases as the inverse square of the distance, could be changed a little by supposing that the force decreases with some other power of the distance, close to but not precisely the same as the inverse square. To test Newton's theory experimentally we might compare observations of the solar system with what would be expected for a force that falls off as some unknown power of the distance and in that way put a limit on how far from an inverse square this power of distance can be. Even general relativity could be changed a little, for instance by including more complicated small terms in the field equations or by introducing weakly interacting new fields into the theory. It is striking that it has so far not been possible to find a logically consistent theory that is close to quantum mechanics, other than quantum mechanics itself.

I tried to construct such a theory a few years ago. My purpose was not seriously to propose an alternative to quantum mechanics but only to have *some* theory whose predictions would be close to but not quite the same as those of quantum mechanics, to serve as a foil that might be tested experimentally. I was trying in this way to give experimental physicists an idea of the sort of experiment that might provide interesting quantitative tests of the validity of quantum mechanics. One wants to test quantum mechanics itself, and not any particular quantum-mechanical theory like the standard model, so to distinguish experimentally between quantum mechanics and its alternatives one must check some very general feature of any possible quantum-mechanical theory. In inventing an alternative to quantum mechanics I fastened on the one general feature of quantum mechanics that has always seemed somewhat more arbitrary than the others, its *linearity.*

I need to say a bit here about the meaning of linearity. Recall that the values of the wave function of any system change at rates that depend on these values, as well as on the nature of the system and its environment. For instance, the rate of change of the *here* value of our mythical particle's wave function is a constant number times the *here* value, plus another constant number times the *there* value. A dynamical rule of this particular type is called linear, because, if we change one value of the wave function at any one time and plot a graph of any value of the wave function at any later time against the value that has been changed, then, all other things being equal, the graph is a straight line. Speaking very loosely, the response of the system to any change in its state is proportional to that change. One very important consequence of this linearity is that, as Scrooge pointed out, quantum systems cannot exhibit chaos; a small change in the initial conditions produces only a small change in the values of the wave function at any later time.

There are many classical systems that are linear in this sense, but the linearity in classical physics is never exact. Quantum mechanics in contrast is supposed to be exactly linear under all circumstances. If one is going to look for ways of changing quantum mechanics, it is natural to try out the possibility that perhaps the evolution of the wave function is not exactly linear after all.

After some work I came up with a slightly nonlinear alternative to quantum mechanics that seemed to make physical sense and could be easily tested to very high accuracy by checking a general consequence of linearity, that the frequencies of oscillation of any sort of linear system do not depend on how the oscillations are excited. For instance, Galileo noticed that the frequency with which a pendulum goes back and forth does not depend on how far the pendulum swings. This is because, as long as the magnitude of the oscillation is small enough, the pendulum is a linear system; the rates of change of its displacement and momentum are proportional to its momentum and its displacement, respectively. All clocks are based on this feature of oscillations of linear systems, whether pendulums or springs or quartz crystals. A few years ago after a conversation with David Wineland of the National Bureau of Standards, I realized that the spinning nuclei that were used by the bureau to set time standards provided a wonderful test of the linearity of quantum mechanics; in my slightly nonlinear alternative to quantum mechanics the frequency with which the spin axis of the nucleus precesses around a magnetic field would depend very weakly on the angle between the spin axis and the magnetic field. The fact that no such effect had been seen at the Bureau of Standards told me immediately that any nonlinear effects in the nucleus studied (an isotope of beryllium) could contribute no more than one part in a billion billion billion to the energy of the nucleus. Since this work, Wineland and several other experimentalists at Harvard, Princeton, and other laboratories have improved these measurements, so that we now know that nonlinear effects would have to be even smaller than this. The linearity of quantum mechanics, if only approximate, is a rather good approximation after all.

None of this was particularly surprising. Even if there are small nonlinear corrections to quantum mechanics, there was no reason to believe that these corrections should be just large enough to show up in the first round of experiments designed to search for them. What I did find disappointing was that this nonlinear alternative to quantum mechanics turned out to have purely theoretical internal difficulties. For one thing, I could not find any way to extend the nonlinear version of quantum mechanics to theories based on Einstein's special theory of relativity. Then, after my work was published, both N. Gisin in Geneva and my colleague Joseph Polchinski at the University of Texas independently pointed out that in the Einstein-Podolsky-Rosen thought experiment mentioned by Tiny Tim, the nonlinearities of the generalized theory *could* be used to send signals instantaneously over large distances, a result forbidden by special relativity. At least for the present I have given up on the problem; I simply do not know how to change quantum mechanics by a small amount without wrecking it altogether.

This theoretical failure to find a plausible alternative to quantum mechanics, even more than the precise experimental verification of linearity, suggests to me that quantum mechanics is the way it is because any small change in quantum mechanics would lead to logical absurdities. If this is true, quantum mechanics may be a permanent part of physics. Indeed, quantum mechanics may survive not merely as an approximation to a deeper truth, in the way that Newton's theory of gravitation survives as an approximation to Einstein's general theory of relativity, but as a precisely valid feature of the final theory.

CHAPTER V

TALES OF THEORY AND EXPERIMENT

As we grow older
The world becomes stranger, the pattern more complicated
Of dead and living. Not the intense moment
Isolated, with no before and after,
But a lifetime burning in every moment.

T. S. Eliot, *East Coker*

Inow want to tell three stories about advances in twentieth-century physics. A curious fact emerges in these tales: time and again physicists have been guided by their sense of beauty not only in developing new theories but even in judging the validity of physical theories once they are developed. It seems that we are learning how to anticipate the beauty of nature at its most fundamental level. Nothing could be more encouraging that we are actually moving toward the discovery of nature's final laws.

My first tale has to do with the general theory of relativity, Einstein's theory of gravitation. Einstein developed this theory in the years 1907–1915 and presented it to the world in a series of papers in 1915–16. Very briefly, in place of Newton's picture of gravitation as an attraction between all massive bodies, general relativity describes gravitation as an effect of the curvature of space-time produced by both matter and energy. By the mid-1920s this revolutionary theory had become generally accepted as the correct theory of gravitation, a position it has kept since then. How did this happen?

Einstein recognized immediately in 1915 that his theory resolved an old conflict between observations of the solar system and Newtonian theory. Since 1859 there had been a difficulty in understanding the orbit of the planet Mercury within the framework of Newton's theory. According to Newton's mechanics and theory of gravitation, if there were nothing in the universe but the sun and a single planet the planet would move in a perfect ellipse around the sun. The orientation of the ellipse—the way its longer and shorter axes point—would never change; it would be as if the planet's orbit were fixed in space. Because the solar system in fact contains a number of different planets that slightly perturb the sun's gravitational field, the elliptical orbits of all the planets actually precess; that is, they swing around slowly in space. In the nineteenth century it became known that the orbit of the planet Mercury changes its orientation by an angle of about 575 seconds in a century. (One degree equals 3,600 seconds.) But Newtonian theory would predict that Mercury's orbit should precess by 532 seconds per century, a difference of 43 seconds per century. Another way of saying this is that if you wait 225,000 years, the elliptical orbit, after having swung once around through 360 degrees, will be back in its original orientation, whereas Newtonian theory would predict that this would take 244,000 years—not a terribly dramatic discrepancy, but nevertheless one that had bothered astronomers for more than half a century. When Einstein in 1915 worked out the consequences of his new theory, he found that it immediately explained the excess precession of 43 seconds per century in the orbit of Mercury. (One of the effects that contributes to this extra precession in Einstein's theory is the extra gravitational field produced by the energy in the gravitational field itself. In Newton's theory gravitation is produced by mass alone, not energy, and there is no such extra gravitational field.) Einstein recalled later that he was beside himself with delight for several days after this success.

After the war astronomers subjected general relativity to a further experimental test, a measurement of the deflection of light rays by the sun during the solar eclipse of 1919. The photons in a ray of light are deflected by gravitational fields in Einstein's theory in much the same way that a comet entering the solar system from a great distance is deflected by the gravitational field of the sun as it loops around the sun and goes back out to interstellar space. Of course the deflection of light is much less than the deflection of a comet because the light is traveling so much faster, just as a fast comet is deflected less than a slow one. If general relativity is correct the deflection of a light ray grazing the sun would be 1.75 seconds, or about 5 ten thousandths of a degree. (Astronomers must wait for an eclipse to measure this deflection because they are looking for a bending of light rays from a distant star as the rays pass close to the sun, and of course it is difficult to see stars near the sun unless the sun's light is blocked out by the moon in an eclipse. So astronomers measure the position on the celestial sphere of several stars six months before the eclipse when the sun is on the other side of the sky, and then they wait six months for the eclipse to occur, and they measure how much the close passage of the starlight to the sun bends the light rays, as shown by a shift in the apparent position of these stars in the sky.) In 1919 British astronomers mounted expeditions to observe a solar eclipse, from a small town in northeast Brazil and an island in the Gulf of Guinea. They found that the deflection of light rays from several stars was, within the experimental uncertainties, equal to what Einstein had predicted. General relativity thereby won worldwide acclaim and became a subject of cocktail party conversation everywhere.

So isn't it obvious how general relativity supplanted Newton's theory of gravitation? General relativity explained one old anomaly, the excess precession of Mercury's orbit, and it successfully predicted a striking new effect, the deflection of light by the sun. What more need be said?

The anomaly in the orbit of Mercury and the deflection of light were of course part of the story, and an important part at that. But, like everything in the history of science (or I suppose in the history of anything else), the simplicity of the story dissolves when we look at it more closely.

Consider the conflict between Newton's theory and the observed motion of Mercury. Even without general relativity, didn't this show clearly that something was wrong with Newton's theory of gravity? Not necessarily. Any theory like Newton's theory of gravitation that has an enormous scope of application is always plagued by experimental anomalies. There is no theory that is not contradicted by some experiment. Newton's theory of the solar system was contradicted by various astronomical observations continually through its history. By 1916, these discrepancies included not only the anomaly in Mercury's orbit but also anomalies in the motion of Halley's and Encke's comets and in the motion of the moon. All these showed behavior that did not fit Newton's theory. We now know that the explanation of the anomalies in the motion of the comets and the moon had nothing at all to do with the fundamentals of the theory of gravitation. Halley's and Encke's comets do not behave as had been expected from calculations using Newton's theory because in these calculations one did not know the correct way to take into account the pressure exerted by gases escaping the rotating comets when the comets are heated as they pass close to the sun. And, similarly, the motion of the moon is very complicated because the moon is a rather large object and therefore subject to all sorts of complicated tidal forces. With hindsight, it is not surprising that there appeared to be discrepancies in the application of Newton's theory to these phenomena. Similarly, there were several suggestions of how the anomaly in the motion of Mercury could be explained within Newtonian theory. One possibility that was being taken seriously at the beginning of this century was that there might be some kind of matter between Mercury and the sun that causes a slight perturbation to the sun's gravitational field. There is nothing in any single disagreement between theory and experiment that stands up and waves a flag and says, "I am an important anomaly." There was no sure way that a scientist looking critically at the data in the latter part of the nineteenth or the first decade of the twentieth century could have concluded that there was anything important about any of these solar system anomalies. It took theory to explain which were the important observations.

Once Einstein had calculated in 1915 that general relativity entailed an excess precession of Mercury's orbit equal to the observed value of 43 seconds per century, this of course became an important piece of evidence for his theory. In fact, as I will argue later, it could have been taken even more seriously than it was. Perhaps it was the variety of other possible perturbations of Mercury's orbit or perhaps it was a prejudice against validating theories by preexisting data or perhaps it was only the war—at any rate, the success of Einstein's explanation of Mercury's precession did not have anything like the impact of the report from the 1919 eclipse expedition that verified Einstein's prediction of the deflection of light by the sun.

So let us now turn to the deflection of light by the sun. After

1919 astronomers went on to check Einstein's prediction in a number of subsequent eclipses. There was one eclipse in 1922 visible in Australia; one in 1929 in Sumatra; one in 1936 in the USSR; and one in 1947 in Brazil. Some of these observations did seem to yield a result for the deflection of light in agreement with Einstein's theory, but several others found a result that seriously disagreed with Einstein's prediction. And, although the 1919 expedition had reported a 10% experimental uncertainty in the deflection on the basis of observations of a dozen stars, and an agreement with Einstein's theory to an accuracy also of about 10%, several of the later eclipse expeditions found that they could not achieve that accuracy, even though they had observed many more stars. It is true that the 1919 eclipse was unusually favorable for this sort of observation. Nevertheless, I am inclined to believe that the astronomers of the 1919 expedition had been carried away with enthusiasm for general relativity in analyzing their data.

Indeed, some scientists at the time had reservations about the 1919 eclipse data. In a report to the 1921 Nobel committee, Svante Arrhenius referred to various criticisms of the reported results on the bending of light. Once in Jerusalem I met an elderly Professor Sambursky, who in 1919 had been a colleague of Einstein at Berlin. He told me that the astronomers and physicists at Berlin had been skeptical that the British astronomers could really have achieved such an accurate test of Einstein's theory.

This is not for a moment to suggest that any dishonesty had crept into these observations. You can imagine all the uncertainties that plague you when you measure the deflection of light by the sun. You are looking at a star that appears in the sky close to the sun's disk when the sun is blotted out by the moon. You are comparing the position of the star on photographic plates at two times six months apart. The telescope may have been focused differently in the two observations. The photographic plate itself may have expanded or shrunk in the interval. And so on. As in all experiments, all sorts of corrections are needed. The astronomer makes these corrections the best way that he or she can. But, if one knows the answer, there is a natural tendency to keep on making these corrections until one has the "right" answer and then to stop looking for further corrections. Indeed, the astronomers of the 1919 eclipse expedition were accused of bias in throwing out the data from one of the photographic plates that would have been in conflict with Einstein's prediction, a result they blamed on a change of focus of the telescope. With hindsight we can say that the British astronomers were right, but I would not be surprised if they had gone on finding corrections until finally their result with all these corrections fit Einstein's theory.

It is widely supposed that the true test of a theory is in the comparison of its predictions with the results of experiment. Yet, with the benefit of hindsight, one can say today that Einstein's successful explanation in 1915 of the previously measured anomaly in Mercury's orbit was a far more solid test of general relativity than the verification of his calculation of the deflection of light by the sun in observations of the eclipse of 1919 or in later eclipses. That is, in the case of general relativity a *retrodiction*, the calculation of the already-known anomalous motion of Mercury, in fact provided a more reliable test of the theory than a true *prediction* of a new effect, the deflection of light by gravitational fields.

I think that people emphasize prediction in validating scientific theories because the classic attitude of commentators on science is not to trust the theorist. The fear is that the theorist adjusts his or her theory to fit whatever experimental facts are already known, so that for the theory to fit these facts is not a reliable test of the theory.

But even though Einstein had learned of the excess precession of Mercury's orbit as early as 1907, no one who knows anything about how general relativity was developed by Einstein, who at all follows Einstein's logic, could possibly think that Einstein developed general relativity in order to explain this precession. (I come back in a minute to Einstein's actual train of thought.) Often it is a successful *prediction* that one should really distrust. In the case of a true prediction, like Einstein's prediction of the bending of light by the sun, it is true that the theorist does not know the experimental result when she develops the theory, but on the other hand the experimentalist does know about the theoretical result when he does the experiment. And that can lead, and historically has led, to as many wrong turns as overreliance on successful retrodictions. I repeat: it is not that experimentalists falsify their data. To the best of my knowledge there never has been an important case of outright falsification of data in physics. But experimentalists who know the result that they are theoretically supposed to get naturally find it difficult to stop looking for observational errors when they do not get that result or to go on looking for errors when they do. It is a testimonial to the strength of character of experimentalists that they do not always get the results they expect.

To summarize the story so far, we have seen that the early experimental evidence for general relativity boiled down to a single successful retrodiction, of the anomaly in Mercury's motion, which probably was not taken as seriously as it deserved, plus a prediction of a new effect, the deflection of light by the sun, whose apparent success did have a huge impact, but was in fact not as conclusive as was generally supposed at the time and was greeted with skepticism by at least a few scientists. It was not until after World War II that new techniques of radar and radio astronomy led to a significant improvement in the accuracy of these experimental tests of general relativity. We can now say that the predictions of general relativity for the deflection (and also delay) of light passing the sun and for the orbital motion not only of Mercury but also of the asteroid Icarus and other natural and man-made bodies have been confirmed with experimental uncertainties less than 1%. But this was a long time in coming.

Nevertheless, despite the weakness of the early experimental evidence for general relativity, Einstein's theory became the standard textbook theory of gravitation in the 1920s and retained that position from then on, even while the various eclipse expeditions of the 1920s and 1930s were reporting at best equivocal evidence for the theory. I remember that, when I learned general relativity in the 1950s, before modern radar and radio astronomy began to give impressive new evidence for the theory, I took it for granted that general relativity was more or less correct. Perhaps all of us were just gullible and lucky, but I do not think that is the real explanation. I believe that the general acceptance of general relativity was due in large part to the attractions of the theory itself—in short, to its beauty.

Einstein in developing general relativity had pursued a line of thought that could be followed by the subsequent generations of physicists who would set out to learn the theory and that would exert over them the same seductive qualities that had attracted Einstein in the first place. We can trace the story back to 1905, Einstein's *annus mirabilis*. In that year, while also working out the quantum theory of light and a theory of the motion of small particles in fluids, Einstein developed a new view of space and time, now called the special theory of relativity. This theory fit in well with the accepted theory of electricity and magnetism, Maxwell's electrodynamics. An observer moving with constant speed would observe space and time intervals and electric magnetic fields to be modified by the observer's motion in just such a way that Maxwell's equations would still be valid despite the motion (not surprising, because special relativity was developed specifically to satisfy this requirement). But special relativity did not fit at all well with the Newtonian theory of gravitation. For one thing, in Newton's theory the gravitational force between the sun and a planet depends on the distance between their positions *at the same time,* but in special relativity there is no absolute meaning to simultaneity—depending on their state of motion, different observers will disagree as to whether one event occurs before or after or at the same time as another event.

There were several ways that Newton's theory could have been patched up so that it would be in accord with special relativity, and Einstein tried at least one of them before he came to general relativity. The clue that in 1907 started him on the path to general relativity was a familiar and distinctive property of gravitation: the force of gravity is proportional to the mass of the body on which it acts. Einstein reflected that this is just like the so-called inertial forces that act on us when we move with a non-uniform speed or direction. It is an inertial force that pushes passengers back in their seats when an airplane accelerates down the runway. The centrifugal force that keeps the earth from falling into the sun is also an inertial force. All these inertial forces are, like gravitational forces, proportional to the mass of the body on which they act. We on earth do not feel either the gravitational field of the sun or the centrifugal force caused by the earth's motion around the sun because the two forces balance each other, but this balance would be spoiled if one force was proportional to the mass of the objects on which it acts and the other was not; some objects might then fall off the earth into the sun and others could be thrown off the earth into interstellar space. In general the fact that gravitational and inertial forces are both proportional to the mass of the body on which they act but depend on no other property of the body makes it possible at any point in any gravitational field to identify a "freely falling frame of reference" in which neither gravitational nor inertial forces are felt because they are in perfect balance for all bodies. When we do feel gravitational or inertial forces it is because we are not in a freely falling frame. For example, on the earth's surface freely falling bodies accelerate toward the center of the earth at 32 feet per second per second, and we feel a gravitational force unless we happen to be accelerating downward at the same rate. Einstein made a logical jump and guessed that gravitational and inertial forces were at bottom the same thing. He called this the principle of equivalence of gravitation and inertia, or the equivalence principle for short. According to this principle, any gravitational field is completely described by telling which frame of reference is freely falling at each point in space and time.

Einstein spent almost a decade after 1907 searching for an appropriate mathematical framework for these ideas. Finally he found just what he needed in a profound analogy between the role of gravitation in physics and that of curvature in geometry. The fact that the force of gravity can be made to disappear for a brief time over a small region around any point in a gravitational field by adopting a suitable freely falling frame of reference is just like the property of curved surfaces, that we can make a map that despite the curvature of the surface correctly indicates distances and directions in the immediate neighborhood of any point we like. If the surface is curved, no one map will correctly indicate distances and directions everywhere; any map of a large region is a compromise, distorting distances and directions in one way or another. The familiar Mercator projection used in maps of the earth gives a good idea of distances and directions near the equator, but produces horrible distortions near the poles, with Greenland swelling to many times its actual size. In the same way, it is one sign of being in a gravitational field that there is no *one* freely falling frame of reference in which gravitational and inertial effects cancel everywhere.

Starting with this analogy between gravitation and curvature, Einstein leaped to the conclusion that gravitation is nothing more or less than an effect of the curvature of space and time. To implement this idea he needed a mathematical theory of curved spaces that went beyond the familiar geometry of the spherical two-dimensional surface of the earth. Einstein was the greatest physicist the world has seen since Newton, and he knew as much mathematics as most physicists of his time, but he was not himself a mathematician. Finally he found just what he needed ready-made in a theory of curved space that had been worked out by Riemann and other mathematicians in the previous century. In its final form, the general theory of relativity was just a reinterpretation of the existing mathematics of curved spaces in terms of gravitation, together with a *field equation* that specified the curvature produced by any given amount of matter and energy. Remarkably, for the small densities and low velocities of the solar system, general relativity gave just the same results as Newton's theory of gravitation, with the two theories distinguished only by tiny effects like the precession of orbits and the deflection of light.

I have more to say later about the beauty of the general theory of relativity. For the present, I hope that I have said enough to give you some feeling for the attractiveness of these ideas. I believe that it was this intrinsic attractiveness that preserved physicists' belief in general relativity during the decades when the evidence from successive eclipse expeditions continued to prove so disappointing.

This impression is reinforced when we look to the reception of general relativity in its first few years, *before* the 1919 eclipse expedition. Most important of all was the reception of general relativity by Einstein himself. In a postcard to the older theorist Arnold Sommerfeld dated February 8, 1916, three years before the eclipse expedition, Einstein commented, "Of the general theory of relativity you will be convinced, once you have studied it. Therefore I am not going to defend it with a single word." I have no way of knowing to what extent the successful calculation of the precession of Mercury's orbit contributed to Einstein's confidence in general relativity in 1916, but well before then, before he did this calculation, something must have given him enough confidence in the ideas that underlie general relativity to keep him working on it, and this could only have been the attractiveness of the ideas themselves.

We should not undervalue this early confidence. The history of science offers countless examples of scientists who had good ideas that they did not at the time pursue, even though years later these ideas were found (often by others) to lead to important progress. It is a common error to suppose that scientists are necessarily devoted advocates of their own ideas. Very often the scientist who first conceives a new idea subjects it to unfounded or excessive criticism because he or she would have to work long and hard and (more important) give up other research if this idea were to be seriously pursued.

As it happened, physicists *were* impressed by general relativity. A band of cognoscenti in Germany and elsewhere heard of general relativity and regarded it as promising and important well before the 1919 eclipse expedition. These included not only Sommerfeld in Munich, Max Born and David Hilbert in Göttingen, and Hendrik Lorentz in Leiden, with all of whom Einstein had been in touch during the war, but also Paul Langevin in France and Arthur Eddington in England, who instigated the 1919 eclipse expedition. The nominations of Einstein for Nobel prizes from 1916 on are instructive. In 1916 Felix Ehrenhaft nominated him for his theory of Brownian motion and for special and general relativity. In 1917 A. Haas nominated him for general relativity (quoting as evidence the successful calculation of the precession of Mercury's orbit). Also in 1917 Emil Warburg nominated Einstein for a variety of contributions including general relativity. More nominations along similar lines were received in 1918. Then in 1919, four months before the eclipse expedition, Max Planck, one of the fathers of modern physics, nominated Einstein for general relativity, and commented that Einstein "made the first step beyond Newton."

I do not mean to imply that the world community of physicists was unanimously and unreservedly convinced of the validity of general relativity from the beginning. For instance, the report of the 1919 Nobel committee suggested waiting for the eclipse of May 29 before reaching a decision about general relativity, and even after 1919, when Einstein in 1921 was finally awarded the Nobel Prize, it was not explicitly for special or general relativity, but "for his services to theoretical physics, and especially for his discovery of the law of the photoelectric effect."

It is not really so important to pinpoint the moment when physicists became 75% or 90% or 99% convinced of the correctness of general relativity. The important thing for the progress of physics is not the decision that a theory is true, but the decision that it is worth taking seriously—worth teaching to graduate students, worth writing textbooks about, above all, worth incorporating into one's own research. From this point of view, the most crucial early converts won by general relativity (after Einstein himself) were the British astronomers, who became convinced not that general relativity was true but that it was plausible enough and beautiful enough to be worth devoting a fair fraction of their own research careers to test its predictions, and who traveled thousands of miles from Britain to observe the eclipse of 1919. But even earlier, before general relativity was completed and before the successful calculation of the precession of Mercury's orbit, the beauty of Einstein's ideas had led Erwin Freundlich of the Royal Observatory in Berlin to mount a Krupp-financed expedition to the Crimea to observe the eclipse of 1914. (The war forestalled his observations, and Freundlich was briefly interned in Russia for his pains.)

The reception of general relativity depended neither on experimental data alone nor on the intrinsic qualities of the theory alone but on a tangled web of theory and experiment. I have emphasized the theoretical side of this story as a counterweight to a naive overemphasis on experiment. Scientists and historians of science have long ago given up the old view of Francis Bacon, that scientific hypotheses should be developed by patient and unprejudiced observation of nature. It is glaringly obvious that Einstein did not develop general relativity by poring over astronomical data. Nevertheless, one still finds a widespread acceptance of John Stuart Mill's view that it is by observation alone that we can *test* our theories. But, as we have seen here, in the acceptance of general relativity aesthetic judgments and experimental data were inseparably linked.

In one sense there was from the beginning a vast amount of experimental data that supported general relativity—namely, observations of the way the earth goes around the sun and the way the moon goes around the earth and all the other detailed observations of the solar system, going back to Tycho Brahe and before, that had already been explained by Newtonian theory. This may at first seem like a very peculiar sort of evidence. Not only are we now citing as evidence for general relativity a retrodiction, a calculation of planetary motions that had been already measured at the time the theory was developed—we are now speaking of astronomical observations that not only had been performed before Einstein developed his theory but also had been already explained by another theory, that of Newton. How could a successful prediction or retrodiction of such observations be counted as a triumph for general relativity?

To understand this, you must look more closely at both Newton's and Einstein's theories. Newtonian physics did explain virtually all the observed motions of the solar system, but at the cost of introducing a set of somewhat arbitrary assumptions. For example, consider the law that says that the gravitational force produced by any body decreases like the inverse square of the distance from the body. In Newton's theory there is nothing about an inverse-square law that is particularly compelling. Newton developed the idea of an inverse-square law in order to explain known facts about the solar system, like Kepler's relation between the size of planetary orbits and the time it takes planets to go around the sun. Apart from these observational facts, in Newton's theory one could have replaced the inverse-square law with an inverse-cube law or an inverse 2.01-power law without the slightest change in the conceptual framework of the theory. It would be changing a minor detail in the theory. Einstein's theory was far less arbitrary, far more rigid. For slowly moving bodies in weak gravitational fields, for which one can legitimately speak of an ordinary gravitational force, general relativity *requires* that the force must fall off according to an inverse-square law. It is not possible in general relativity to adjust the theory to get anything but an inverse-square law without doing violence to the underlying assumptions of the theory.

Also, as Einstein particularly emphasized in his writing, the fact that the force of gravity on a small object is proportional to the object's mass but depends on no other property of the object appears rather arbitrary in Newton's theory. The gravitational force might in Newton's theory have depended for instance on the size or shape or chemical composition of the body without upsetting the underlying conceptual basis of the theory. In Einstein's theory the force that gravity exerts on any object *must* be both proportional to the object's mass and independent of any other of its properties;* if this were not true, then gravitational and inertial forces would balance in different ways for different bodies, and it would not be possible to talk of a freely falling frame of reference in which no body feels the effects of gravitation. This would rule out the interpretation of gravitation as a geometric effect of the curvature of space-time. So, again, Einstein's theory had a rigidity that Newton's theory lacked, and for this reason Einstein could feel that he had explained the ordinary motions of the solar system in a way that Newton had not.

Unfortunately this notion of the rigidity of physical theories is very difficult to pin down at all precisely. Newton and Einstein both knew about the general features of planetary motion before they formulated their theories, and Einstein knew that he had to have something like an inverse-square law for the gravitational force in order for his theory to reproduce the successes of Newton's. Also, he knew that he had to wind up with a gravitational force proportional to mass. It is only after the fact, looking at the whole theory as finally developed, that we can say that Einstein's theory explained the inverse-square law or the proportionality of gravitational force to mass, but this judgment is a matter of taste and intuition—it is precisely a judgment that Einstein's theory if modified to allow an alternative to the inverse-square law or a non-proportionality of grav-

---

*Strictly speaking, this is only for slowly moving small objects. For a rapidly moving object the force of gravity also depends on the object's momentum. This is why the gravitational field of the sun is able to deflect light rays, which have momentum but no mass.

itational force to mass would be too ugly to bear. So again we carry with us our aesthetic judgments and our whole heritage of theory when we judge the implications of data.

My next tale has to do with quantum electrodynamics—the quantum-mechanical theory of electrons and light. It is in a way the mirror image of the first tale. For forty years general relativity was widely accepted as the correct theory of gravitation despite the slimness of the evidence for it, because the theory was compellingly beautiful. On the other hand quantum electrodynamics was very early supported by a great wealth of experimental data but nevertheless was viewed with distrust for twenty years because of an internal theoretical contradiction that it seemed could only be resolved in ugly ways.

Quantum mechanics was applied to electric and magnetic fields in one of the first papers on quantum mechanics, the *Dreimännerarbeit* of Max Born, Werner Heisenberg, and Pascual Jordan in 1926. They were able to calculate that the energy and momentum of the electric and magnetic fields in a ray of light come in bundles that behave as particles, and thus to justify Einstein's introduction in 1905 of the particles of light known as photons. The other main ingredient of quantum electrodynamics was supplied by Paul Dirac in 1928. In its original form, Dirac's theory showed how to make the quantum-mechanical description of electrons in terms of wave functions consistent with the special theory of relativity. One of the most important consequences of Dirac's theory was that, for every species of charged particle such as the electron, there must be another species with the same mass but opposite electric charge, known as its antiparticle. The antiparticle of the electron was discovered in 1932 and is today known as the positron. Quantum electrodynamics was used in the late 1920s and early 1930s to calculate a wide variety of physical processes (such as the scattering of a photon colliding with an electron, the scattering of one electron by another, and the annihilation or production of an electron and a positron) with results that were generally in excellent agreement with experiment.

Nevertheless by the mid-1930s it had become standard wisdom that quantum electrodynamics was not to be taken seriously except as an approximation, valid only for reactions involving photons, electrons, and positrons of sufficiently low energy. The problem was not the sort that usually appears in popular histories of science, a conflict between theoretical expectations and experimental data, but rather a persistent internal contradiction within physical theory itself. It was the problem of infinities.

This problem was noticed in various forms by Heisenberg and Pauli and by the Swedish physicist Ivar Waller, but it appeared most clearly and disturbingly in a 1930 article by the young American theorist, Julius Robert Oppenheimer. Oppenheimer was trying to use quantum electrodynamics to calculate a subtle effect on the energies of atoms. An electron in an atom may emit a particle of light, a photon, continue in its orbit for a while, and then reabsorb the photon, like a football quarterback catching his own forward pass. The photon never gets outside the atom and makes its presence known only indirectly, through its effects on properties of the atom like its energy and its magnetic field. (Such photons are called *virtual*.) According to the rules of quantum electrodynamics, this process produces a shift in the energy of the atomic state that may be calculated by adding up an infinite number of contributions, one contribution for each possible value of the energy that can be given to the virtual photon, with no limit placed on the photon's energy. Oppenheimer found in his calculation that, because the sum includes contributions from photons of unlimitedly high energy, it turns out to be infinite, leading to an infinite shift in the energy of the atom.* High energy corresponds to short wavelengths; because ultraviolet light has a wavelength shorter than that of visible light, this infinity became known as the ultraviolet catastrophe.

Throughout the 1930s and early 1940s there was a consensus among physicists that the appearance of the ultraviolet catastrophe in Oppenheimer's calculation and similar calculations showed that the existing theory of electrons and photons simply could not be trusted for particles with energies above a few million volts. Oppenheimer himself was the foremost proponent of this view. This was in part because Oppenheimer was a leader in the study of cosmic rays, the high-energy particles that plow into the earth's atmosphere from outer space, and his study of the way that these cosmic-ray particles interact with the atmosphere indicated that something funny was going on for particles of high energy. There was indeed something funny going on, but in fact it had nothing to do with any breakdown in the quantum theory of electrons and photons; it was rather a sign of the production of particles of new types, the particles that are today called muons. But even after all this was cleared up by the discovery of muons in 1937, it remained the conventional wisdom that something goes wrong when quantum electrodynamics is applied to electrons and photons of high energy.

The problem of infinities could have been solved by brute force, by simply decreeing that electrons can emit and absorb photons only with energies below some limiting value. All the successes scored in the 1930s by quantum electrodynamics in explaining the interactions of electrons and photons involved

---

*Not every sum of an infinite number of things is infinite. E.g., although the sum $1 + \frac{1}{2} + \frac{1}{3} + \frac{1}{4} + \ldots$ *is* infinite, the sum $1 + \frac{1}{2} + \frac{1}{4} + \frac{1}{8} + \ldots$ turns out to have the perfectly finite value 2.

low-energy photons, so these successes could be preserved by supposing that this limiting value of photon energies is sufficiently high, for example, 10 million volts. With this sort of limit on virtual photon energies, quantum electrodynamics would predict a very small energy shift of atoms. No one at this time had measured the energies of atoms with enough precision to tell whether or not this tiny energy shift was actually present, so there was no question of any disagreement with experiment. (In fact, quantum electrodynamics was regarded with such pessimism that no one even tried to calculate just what the energy shift would be.) The trouble with this solution to the problem of infinities was not that it was in conflict with experiment but that it was too arbitrary and ugly to contemplate.

In the physics literature of the 1930s and 1940s one can find a host of other possible unpalatable solutions of the problem of infinities, including even theories in which the infinity caused by emitting and reabsorbing high-energy photons is canceled by other processes of negative probability. The concept of negative probability is of course meaningless; its introduction into physics is a measure of the desperation that was felt over the problem of infinities.

In the end, the solution to the problem of infinities that emerged in the late 1940s was much more natural and less revolutionary. The problem famously came to a head in the beginning of June 1947 at a conference at the Ram's Head Inn on Shelter Island, off the coast of Long Island. The conference had been organized to bring together physicists who were ready after the war years to start thinking again about the fundamental problems of physics. It turned out to be the most important physics conference since the Solvay Conference in Brussels where Einstein and Bohr had battled over the future of quantum mechanics fifteen years earlier.

Among the physicists at Shelter Island was Willis Lamb, a

young physicist at Columbia University. Using some of the microwave radar technology developed during the war, Lamb had just succeeded in measuring precisely the sort of effect that Oppenheimer had tried to calculate in 1930, a shift in the energy of the hydrogen atom owing to photo emissions and reabsorptions. This shift has since become known as the Lamb shift. This measurement in itself did nothing to solve the problem of infinities but forced physicists to come to grips with this problem again in order to account for the measured value of the Lamb shift. The solution they found was to govern the course of physics ever since.

Several of the theorists at the Shelter Island conference had already heard of Lamb's result and had come to the meeting armed with an idea about how the Lamb shift could be calculated using the principles of quantum electrodynamics despite the problem of infinities. They reasoned that the shift in the energy of an atom owing to emission and reabsorption of photons is not really an observable; the only observable is the total energy of the atom, which is calculated by adding this energy shift to the energy calculated in 1928 by Dirac. This total energy depends on the *bare mass* and *bare charge* of the electron, the mass and charge that appear in the equations of the theory before we start worrying about photon emissions and reabsorptions. But free electrons as well as electrons in atoms are always emitting and reabsorbing photons that affect the electron's mass and electric charge, and so the bare mass and charge are not the same as the measured electron mass and charge that are listed in tables of elementary particles. In fact, in order to account for the observed values (which of course are finite) of the mass and charge of the electron, the bare mass and charge must themselves be infinite. The total energy of the atom is thus the sum of two terms, both infinite: the bare energy that is infinite because it depends on the infinite bare mass and charge, and the energy shift calculated by Oppenheimer that is infinite because it receives contributions from virtual photons of unlimited energy. This raised the question: Is it possible that these two infinities cancel and leave a finite total energy?

At first glance, the answer seemed to be a discouraging no. But Oppenheimer had left something out of his calculation. The energy shift receives contributions not only from processes in which an electron emits and then reabsorbs a photon but also from processes in which a positron, a photon, and a second electron appear spontaneously out of empty space, with the photon then being absorbed in the annihilation of the positron and the original electron. In fact this bizarre process *must* be included in the calculation in order for the final answer for the energy of the atom to depend on the atom's velocity in the way dictated by the special theory of relativity. (This is one example of the result found long before by Dirac, that a quantum-mechanical theory of the electron is consistent with special relativity only if the theory also contains the positron, the antiparticle of the electron.) One of the theorists at Shelter Island was Victor Weisskopf, who had already in 1936 calculated the energy shift resulting from this positron process and found that it nearly cancels the infinity found by Oppenheimer.* It was not too hard to guess that, if one took the positron process into account *and* took into account the difference between the bare mass and charge of the electron and their observed values, then the infinities in the energy shifts would be canceled altogether.

Although Oppenheimer and Weisskopf were at the Shelter Island meeting, the theorist there who first calculated the Lamb

---

*To be a little more specific, the inclusion of this positron process made the sum over energies behave like the series $1 + \frac{1}{2} + \frac{1}{3} + \ldots$ instead of $1 + 2 + 3 + 4 \ldots$. Both sums are infinite, but one is less infinite than the other, in the sense that it takes less effort to figure out what to do about it.

shift was Hans Bethe, already famous for his work in nuclear physics, including in the 1930s the description of the chains of nuclear reactions that make the stars shine. Building on the ideas that had been buzzing around Shelter Island, Bethe on the train ride back from the conference worked out a rough calculation of the energy shift that Lamb had measured. He still did not have available really effective techniques to include the positrons and other effects of special relativity in this sort of calculation, and his work on the train followed pretty closely Oppenheimer's efforts seventeen years earlier. The difference was that, when Bethe encountered an infinity, he now threw away the contributions to the energy shift from the emission and absorption of photons of high energy (Bethe somewhat arbitrarily took the limit on photon energies to be the energy in the mass of the electron), and so he obtained a finite result, in fair agreement with Lamb's measurement. It was almost a calculation that Oppenheimer could have done in 1930, but it took the urgency of an experiment that needed to be explained and the encouragement of the ideas that were in the air at Shelter Island to get anyone to push the calculation through to completion.

It was not long before physicists did more accurate calculations of the Lamb shift that included positrons and other relativistic effects. The importance of these calculations was not so much that they obtained a more accurate result but that the problem of infinities had been tamed; the infinities turned out to cancel without having arbitrarily to throw away the contributions of high-energy virtual photons.

As Nietzsche says, "What does not kill us, makes us stronger." Quantum electrodynamics had almost been killed off by the problem of infinities but had been saved by the idea of canceling the infinities in a redefinition or *renormalization* of the electron mass and charge. But in order for the problem of infinities to be solved in this way, it is necessary that the infinities occur in calculations in only certain very limited ways, which is the case only for a limited class of specially simple quantum field theories. Such theories are called *renormalizable*. The simplest version of quantum electrodynamics is renormalizable in this sense, but any sort of small change in this theory would spoil this property and lead to a theory with infinities that could not be canceled by a redefinition of the constants of the theory. Thus this theory was not only mathematically satisfactory and in agreement with experiment, but it seemed to contain within itself an explanation of why it was the way it was; any small change in the theory would lead not only to a disagreement with experiment but to results that were totally absurd—infinite answers to perfectly sensible questions.

The 1948 calculations of the Lamb shift were still terribly complicated because, although these calculations now included positrons, they gave the Lamb shift as a sum of terms that individually violated the special theory of relativity, with only the final answer consistent with the theory of relativity. Meanwhile Richard Feynman, Julian Schwinger, and Sin-itiro Tomonaga were independently developing much simpler methods of calculation that are consistent with the theory of relativity at every step. They used these techniques to do other calculations, some of them in spectacular agreement with experiment. For instance, the electron has a tiny magnetic field, originally calculated in 1928 by Dirac on the basis of his relativistic quantum theory of the electron. Right after the Shelter Island conference Schwinger published the results of an approximate calculation of the shift in the strength of the electron's magnetic field that is caused by processes in which virtual photons are emitted and reabsorbed. This calculation has been continually refined since then, with the modern result that the magnetic field of the electron is increased by photon emissions and reabsorptions and similar effects by a factor 1.00115965214 (with an uncertainty of about 3 in the last digit) over the old Dirac prediction, in which these photon emissions and reabsorptions had been ignored. At just about the time that Schwinger was doing his calculation, experiments by I. I. Rabi and his group at Columbia were revealing that the electron's magnetic field is indeed a little larger than the old Dirac value, and by just about the amount calculated by Schwinger. A recent experimental result is that the magnetic field of the electron is greater than the Dirac value by a factor 1.001159652188, with an uncertainty of about 4 in the last digit. The numerical agreement between theory and experiment here is perhaps the most impressive in all science.

After such successes, it is not surprising that quantum electrodynamics in its simple renormalizable version has become generally accepted as the correct theory of photons and electrons. Nevertheless, despite the experimental success of the theory, and even though the infinities in this theory all cancel when one handles them correctly, the fact that the infinities occur at all continues to produce grumbling about quantum electrodynamics and similar theories. Dirac in particular always referred to renormalization as sweeping the infinities under the rug. I disagreed with Dirac and argued the point with him at conferences at Coral Gables and Lake Constance. Taking account of the difference between the bare charge and mass of the electron and their measured values is not merely a trick that is invented to get rid of infinities; it is something we would have to do even if everything was finite. There is nothing arbitrary or ad hoc about the procedure; it is simply a matter of correctly identifying what we are actually measuring in laboratory measurements of the electron's mass and charge. I did not see what was so terrible about an infinity in the bare mass and charge as long as the final answers for physical quantities turn out to be finite and unambiguous and in agreement with experiment. It seemed to me that a theory that is as spectacularly successful as quantum electrodynamics has to be more or less correct, although we may not be formulating it in just the right way. But Dirac was unmoved by these arguments. I do not agree with his attitude toward quantum electrodynamics, but I do not think that he was just being stubborn; the demand for a completely finite theory is similar to a host of other aesthetic judgments that theoretical physicists always need to make.

My third tale has to do with the development and final acceptance of the modern theory of the weak nuclear force. This force is not as important in everyday life as electric or magnetic or gravitational forces, but it plays an essential role in the chains of nuclear reactions that generate energy and produce the various chemical elements in the cores of stars.

The weak nuclear force first turned up in the discovery of radioactivity by Henri Becquerel in 1896. In the 1930s it became understood that in the particular kind of radioactivity that was discovered by Becquerel, known as beta decay, the weak nuclear force causes a neutron inside the nucleus to turn into a proton, at the same time creating an electron and another particle known today as an antineutrino, and spitting them out of the nucleus. This is something that is not allowed to happen through any other kind of force. The strong nuclear force that holds the protons and neutrons together in the nucleus and the electromagnetic force that tries to push the protons in the nucleus apart cannot change the identities of these particles, and the gravitational force certainly does not do anything of the sort, so the observation of neutrons changing into protons or protons into neutrons provided evidence of a new kind of force in nature. As its name implies, the weak nuclear force is weaker than the electromagnetic or the strong nuclear forces. This is shown for instance by the fact that nuclear beta decay is so slow; the fastest nuclear beta decays take on the average about a hundredth of a second, languorously slow compared with the typical time scale of processes caused by the strong nuclear force, which is roughly a millionth millionth millionth millionth of a second.

In 1933 Enrico Fermi took the first significant step toward a theory of this new force. In his theory the weak nuclear force does not act at a distance like gravitational or electric or magnetic forces; rather, it converts a neutron into a proton and creates an electron and antineutrino instantaneously, all at the same point in space. There followed a quarter century of experimental effort aimed at tying up the loose ends of the Fermi theory. Chief among these loose ends was the question of how the weak force depends on the relative orientation of the spins of the participating particles. In 1957 this was settled and the Fermi theory of the weak nuclear force was put into its final form.

After this breakthrough of 1957 one could say that there were no anomalies in our understanding of the weak nuclear force. Nevertheless, even though we had a theory that was capable of accounting for everything that was known experimentally about the weak force, physicists in general found the theory highly unsatisfactory, and many of us worked hard, trying to clean up the theory and make it make sense.

The things that were wrong with the Fermi theory were not experimental but theoretical. In the first place, although the theory worked well for nuclear beta decay, when the theory was applied to more exotic processes it gave nonsensical results. Theorists would ask perfectly sensible questions, like what is the probability of the scattering of a neutrino by an antineutrino with which it collides, and when they did the calculation (taking account of the emission and reabsorption of a neutron and an antiproton) the answer would turn out to be infinite.

Such experiments were not being done, you understand, but the calculation gave results that could not possibly agree with any experimental result. As we have seen, infinities like these had been encountered in the theory of electromagnetic forces by Oppenheimer and others in the early 1930s, but in the late 1940s theorists had found that all these infinities in quantum electrodynamics would cancel when the mass and electric charge of the electron are properly defined, or "renormalized." As more and more became known about the weak forces it became increasingly clear that the infinities in Fermi's theory of the weak forces would not cancel in this way; the theory was not renormalizable.

The other thing that was wrong with the theory of weak forces was that it had a large number of arbitrary elements. The basic form of the weak force had been inferred more or less directly from experiment and could have been quite different without violating any known physical principles.

I had worked on the theory of weak forces off and on since graduate school, but in 1967 I was working instead on the strong nuclear forces, the forces that hold neutrons and protons together inside atomic nuclei. I was trying to develop a theory of the strong forces based on an analogy with quantum electrodynamics. I thought that the difference between the strong nuclear forces and electromagnetism might be explained by a phenomenon known as *broken symmetry,* which I will explain later. It did not work. I found myself developing a theory that did not look at all like the strong forces as they were known to us experimentally. Then it suddenly occurred to me that these ideas, although they had turned out to be completely useless as far as the strong forces were concerned, provided a mathematical basis for a theory of the weak nuclear force that would do everything that one might want. I could see the possibility of a theory of the weak force analogous to quantum electrodynamics. Just as the electromagnetic force between distant charged particles is caused by the exchange of photons, a weak force would not act all at once at a single point in space (as in the Fermi theory) but would be caused by the exchange of photonlike particles between particles at different positions. These new photonlike particles could not be massless like the photon (for one thing, if massless they would have been discovered long before), but they were introduced into the theory in a way that was so similar to the way that the photon appears in quantum electrodynamics that I thought that the theory might be renormalizable in the same sense as quantum electrodynamics—that is, that the infinities in the theory could be canceled by a redefinition of the masses and other quantities in the theory. Furthermore, the theory would be highly constrained by its underlying principles and would thus avoid a large part of the arbitrariness of previous theories.

I worked out a particular concrete realization of this theory, that is, a particular set of equations that govern the way the particles interacted and that would have the Fermi theory as a low-energy approximation. I found in doing this, although it had not been my idea at all to start with, that it turned out to be a theory not only of the weak forces, based on an analogy with electromagnetism; it turned out to be a unified theory of the weak and electromagnetic forces that showed that they were both just different aspects of what subsequently became called an *electroweak* force. The photon, the fundamental particle whose emission and absorption causes electromagnetic forces, was joined in a tight-knit family group with the other photonlike particles predicted by the theory: electrically charged W particles whose exchange produces the weak force of beta radioactivity, and a neutral particle I called the "Z," about which more later. (W particles were an old story in speculations about the weak forces; the W stands for "weak." I picked the letter Z for their new sibling because the particle has zero electric charge and also because Z is the last letter of the alphabet, and I hoped that this would be the last member of the family.) Essentially the same theory was worked out independently in 1968 by the Pakistani physicist Abdus Salam, working in Trieste. Aspects of this theory had been anticipated in work of Salam and John Ward, and even earlier by my high-school and Cornell classmate Sheldon Glashow.

This unification of weak and electromagnetic forces was all right as far as it went. One always likes to explain more and more things in terms of fewer and fewer ideas, although I certainly had not realized that was where I was going when I started. But in 1967 this theory provided absolutely no explanations of any experimental anomalies in the physics of weak forces. This theory explained no existing experimental information that had not been previously accounted for by the Fermi theory. The new electroweak theory also received almost no attention at first. But I do not think that the theory failed to interest other physicists solely because it lacked experimental support. Equally important was a purely theoretical question about its internal consistency.

Both Salam and I had stated our opinion that this theory would eliminate the problem of infinities in the weak forces. But we were not clever enough to prove this. In 1971 I received a preprint from a young graduate student at the University of Utrecht named Gerard 't Hooft, in which he claimed to show that this theory actually had solved the problem of the infinities: the infinities in calculations of observable quantities would in fact all cancel just as in quantum electrodynamics.

At first I was not convinced by 't Hooft's paper. I had never heard of him, and the paper used a mathematical method developed by Feynman that I had previously distrusted. A little later I heard that the theorist Ben Lee had taken up 't Hooft's ideas and was trying to get the same results using more conventional mathematical methods. I knew Ben Lee and had great respect for him—if he took 't Hooft's work seriously, then so would I. (Ben later became my best friend and collaborator in physics. He was tragically killed in an automobile accident in 1977.) After that I took a more careful look at what 't Hooft had done and saw that he had indeed found the key to showing that the infinities would cancel.

It was after 't Hooft's paper, although there still was not the slightest grain of new experimental support for this theory, that the electroweak theory began to take off as part of the work agenda of physics. This is one case where one can be fairly precise about the level of interest in a scientific theory, because as it happens the Institute for Scientific Information (ISI) has published a compilation of the number of citations to my first paper on the electroweak theory, as an example of how citation analysis is useful in gaining insight about the history of science. The paper was written in 1967. In 1967 it received zero citations. In 1968 and 1969 the paper again received zero citations. (During this period Salam and I were both trying to prove what eventually 't Hooft did prove, that the theory was free of infinities.) In 1970 it received one citation. (I do not know who that was.) In 1971, the year of the 't Hooft paper, my 1967 paper received three citations, one of them by 't Hooft. In 1972, still without any new experimental support, it suddenly received sixty-five citations. In 1973 it received 165 citations, and the number of citations gradually increased until 1980 when it received 330 citations. A recent study by the ISI showed that this paper was the most frequently cited article on elementary particle physics of the previous half century.

The breakthrough that initially made physicists excited about this theory was the realization that it had solved an internal conceptual problem of particle physics, the problem of infinities in the weak nuclear forces. In 1971 and 1972 there was not yet the slightest experimental evidence that this theory was better than the old Fermi theory.

Then the experimental evidence did start to come in. The exchange of the Z particle would lead to a new kind of weak nuclear force known as a *weak neutral current,* that would show up in the scattering of beams of neutrinos by the nuclei of ordinary atoms. (The term "neutral current" is used because these processes do not involve any exchange of electric charge between the nuclei and other particles.) Experiments to look for this sort of neutrino scattering were set in train at CERN (an acronym that has replaced the original name, Centre Européen de Recherches Nucléaires, of the pan-European laboratory at Geneva) and at Fermilab (outside Chicago). Considerable instigation was needed. Each experiment involved the services of thirty to forty physicists. You do not easily do that kind of experiment unless you have a good idea of what you are trying to accomplish. The discovery of weak neutral currents was first announced in 1973 at CERN, and after some hesitation also at Fermilab. After 1974, when both Fermilab and CERN were in agreement about the existence of the neutral currents, the scientific world became generally convinced that the electroweak theory was correct. The Stockholm newspaper *Dagens Nyheder* even announced in 1975 that Salam and I would win the Nobel Prize for physics that year. (We did not.)

One may ask why the acceptance of the validity of the electroweak theory was so rapid and widespread. Well, of course, the neutral currents had been predicted, and then they were found. Isn't that the way that any theory becomes established? I do not think that one can look at it so simply.

In the first place, neutral currents were nothing new in speculations about weak forces. I once traced the theory of neutral currents back to a 1937 paper by George Gamow and Edward Teller that predicted the existence of weak neutral currents on fairly plausible grounds. There was even previous experimental evidence for neutral currents in the 1960s, but it was never believed; the experimentalists who found evidence for these weak forces always reported it as "background." One new thing in 1973 that was of special importance to experimentalists was a prediction that the strength of the neutral current forces had to lie in a certain range. For instance, in one sort of neutrino reaction they would produce effects 15% to 25% as intense as the ordinary weak forces. This prediction provided a guide to the sensitivity that would be needed in an experimental search for these forces. But what really made 1973 different was that a theory had come along that had the kind of compelling quality, the internal consistency and rigidity, that made it reasonable for physicists to believe they would make more progress in their own scientific work by believing the theory to be true than by waiting for it to go away.

In one sense, the electroweak theory did enjoy experimental support even before the discovery of neutral currents because it had correctly "retrodicted" all the properties of weak forces that had been previously explained by Fermi's theory, as well as all the properties of electromagnetic forces that had been earlier described by quantum electrodynamics. Here again, as in the case of general relativity, one may ask why a retrodiction is to be counted as a success when what is explained had already been explained by an earlier theory? Fermi's theory had explained the features of weak forces by invoking a certain number of arbitrary elements, arbitrary in the same sense that the inverse-square law was arbitrary in Newton's theory of gravitation. The electroweak theory explained these elements (like the dependence of the weak forces on the spins of the participating particles) in a compelling way. But it is not possible to be precise about such judgments; it is a matter of taste and experience.

Suddenly in 1976, three years after the discovery of the neutral currents, a crisis occurred. There was no longer any doubt about the existence of the neutral currents, but experiments in 1976 indicated that these forces did not have some of the properties that the theory predicted. The anomaly emerged in experiments at both Seattle and Oxford on the propagation of polarized light passing through bismuth vapor. It has been known since the work of Jean-Baptiste Biot in 1815 that polarized light passing through solutions of certain sugars experiences a rotation of its polarization to the right or left. For instance, the polarization of light rotates to the right when passing through solutions of the common sugar D-glucose, and to the left when passing through solutions of L-glucose. This is because a molecule of D-glucose is not the same as its mirror image, a molecule of L-glucose, just as a left-handed glove is different from a right-handed glove (in contrast to a hat or a tie, which looks the same whether looked at directly or in a mirror.) One would not normally expect this sort of rotation for polarized light passing through a gas of single atoms like those of bismuth. But the electroweak theory predicted an asymmetry between left and right in the weak nuclear force between electrons and atomic nuclei, caused by the exchange of Z particles, which would give such atoms a kind of "handedness" like a glove or a sugar molecule. (This effect was expected to be particularly large in bismuth because of a peculiarity in its atomic energy levels.) Calculations showed that the asymmetry between left and right in the bismuth atom would cause the polarization of the light passing through bismuth vapor to rotate slowly to the left. To their surprise, the experimentalists at Oxford and Seattle could find no such rotation, and they reported that, if there were any such rotation, it would have to be much slower than had been predicted.

This was really quite a bombshell. These experiments seemed to show that the particular theory that Salam and I had each worked out in 1967–68 could not possibly be correct in its details. But I was not prepared to abandon the general ideas of the electroweak theory. Ever since 't Hooft's 1971 paper I had been quite convinced of the correctness of the outlines of this theory, but I regarded the particular version of this theory that Salam and I had constructed as only one specially simple possibility. For instance, there might be other members of the family formed by the photon and the W and Z particles, or other particles related to the electron and neutrino. Pierre Duhem and W. Van Quine pointed out long ago that a scientific theory can never be absolutely ruled out by experimental data because there is always some way of manipulating the theory or the auxiliary assumptions to create an agreement between theory and experiment. At some point one simply has to decide whether the elaborations that are needed to avoid conflict with experiment are just too ugly to believe.

Indeed, after the Oxford-Seattle experiments many of us theorists went to work to try to find some little modification of the electroweak theory that would explain why the neutral current forces did not have the expected kind of asymmetry between right and left. We thought at first it would be possible to make the theory just a little bit uglier and get it to agree with all the data. I recall that at one point Ben Lee flew to Palo Alto where I was spending the year, and I gave up a long-planned trip to Yosemite to work with him trying to modify the electroweak theory to fit the latest data (including misleading hints of other discrepancies from high-energy neutrino reactions). But nothing seemed to work.

One of the problems was that experiments at CERN and Fermilab had already provided us with a great deal of data about the scattering of neutrinos in collisions with protons and neutrons, almost all of which seemed to verify the original version of the electroweak theory. It was difficult to see how any other theory could do this and also agree with the bismuth results in a natural way—that is, without having to introduce many complications that were carefully adjusted to fit the data. Back at Harvard a little later Howard Georgi and I developed a general argument that there was no natural way to make the electroweak theory agree with the data that were coming from Oxford and Seattle and also with the older data on neutrino reactions. This of course did not stop some theorists from constructing very unnatural theories (an activity that came to be known around Boston as committing an unnatural act), in accordance with the oldest rule of progress in science, that it is better to be doing something than nothing.

Then in 1978, a new experiment at Stanford measured the weak force between electrons and atomic nuclei in an entirely different way, not by using the electrons in the atoms of bismuth but by scattering a beam of electrons from the high-energy accelerator at Stanford off the nuclei of deuterium. (There was nothing special about the choice of deuterium; it was simply a convenient source of protons and neutrons.) The experimenters now found the expected asymmetry between right and left. In this experiment the asymmetry showed up as a difference in the rate of scattering between electrons that are spinning to the left or to the right. (We say that a moving particle is spinning to the right or left if the fingers of the right or left hand point in the direction of the spin when the thumb is pointed in the direction of motion.) The difference in the scattering rates was measured to be about one part in ten thousand, which is just what the theory had predicted.

Suddenly particle physicists everywhere jumped to the conclusion that the original version of the electroweak theory was correct after all. But notice that there were still two experiments that contradicted the theory's predictions for the neutral-current weak force between electrons and nuclei and only one that supported them, and in a rather different context. Why then as soon as that one experiment came along and found agreement with the electroweak theory did physicists generally agree that the theory must indeed be correct? One of the reasons surely was that we were all relieved that we were not going to have to deal with any of the unnatural variants of the original electroweak theory. The aesthetic criterion of naturalness was being used to help physicists weigh conflicting experimental data.

The electroweak theory has continued to be tested experimentally. The Stanford experiment has not been repeated, but several groups of atomic physicists have searched for left-right asymmetries not only in bismuth but in other atoms like thallium and cesium. (Even before the Stanford experiment a group in Novosibirsk had reported seeing the expected asymmetry in bismuth, a report that was not much heeded before the Stanford results, partly because in the West Soviet experimental physics did not have a high reputation for accuracy.) There have been new experiments at Berkeley and Paris, and the physicists in Oxford and Seattle have repeated their experiments. There now is general agreement among experimentalists as well as among theorists that the predicted left-right asymmetry effect really is there with about the expected magnitude in atoms as well as in the high-energy electron scattering studied in the Stanford accelerator experiment. The most dramatic tests of the electroweak theory certainly were the experiments by a group at CERN headed by Carlo Rubbia. In 1983 they discovered the W particles, and in 1984 the Z particle, particles whose existence and properties had been correctly predicted by the electroweak theory in its original version.

Looking back at these events, I feel some regret that I spent so much time in trying to fix up the electroweak theory to make it agree with the Oxford-Seattle data. I wish that I had gone to

Yosemite as planned in 1977; I still have not been there. The whole story is a nice illustration of a half-serious maxim attributed to Eddington: One should never believe any experiment until it has been confirmed by theory.

I do not want to leave you with the impression that this is the way that experiment and theory always affect each other and the progress of science. I have been emphasizing the importance of theory here because I want to counteract a widespread point of view that seems to me overly empiricist. But in fact one can go through the history of important experiments in physics and find many varied roles played by these experiments and many different ways that theory and experiment have interacted. It appears that anything you say about the way that theory and experiment *may* interact is likely to be correct, and anything you say about the way that theory and experiment *must* interact is likely to be wrong.

The search for neutral-current weak forces at CERN and Fermilab provides an example of the class of experiments that are undertaken in order to test theoretical ideas that are not yet generally accepted. These experiments sometimes confirm but sometimes refute the theorist's ideas. A few years ago Frank Wilczek and I independently predicted a new kind of particle. We agreed to call this particle the *axion*, not aware that this was also the name of a brand of detergent. Experimentalists looked for the axion and did not find it—at least not with the properties we had anticipated. The idea either is incorrect or needs modification. I did once receive a message from a group of physicists meeting at Aspen proclaiming, "We found it!" but the message was attached to a box of the detergent.

There are also experiments that present us with complete surprises that no theorist had anticipated. In this category are the experiments that discovered X rays or the so-called strange particles or, for that matter, the anomalous precession of the orbit of the planet Mercury. These I think are the experiments that bring the most joy to the hearts of experimentalists and journalists.

There are also experiments that present us with *almost* complete surprises—that is, that find effects that had been discussed as a possibility, but only as a logical possibility that there was no compelling reason to expect. These include the experiments that discovered the violation of the so-called time-reversal symmetry and the experiments that found certain new particles, such as the "bottom" quark and a sort of very heavy electron known as the tau lepton.

There is also an interesting class of experiments that have found effects that had been predicted by theorists but that were nevertheless discovered accidentally because the experimentalists did not know of the prediction, either because the theorists did not have enough faith in their theory to advertise it to experimentalists or because the channels of scientific communication were just too noisy. Among these experiments are the discovery of a universal background of radio static left over from the big bang and the discovery of the positron.

Then there are experiments that are done even though one knows the answer, even though the theoretical prediction is so firm that the theory is beyond serious doubt, because the phenomena themselves are so entrancing and offer so many possibilities of further experiments that one simply has to go ahead and find these things. I would include in this category the discovery of the antiproton and the neutrino and the more recent discovery of the W and Z particles. Also included here are searches for various exotic effects predicted by general relativity, like gravitational radiation.

Finally one can imagine a category of experiments that *refute* well-accepted theories, theories that have become part of the standard consensus of physics. *Under this category I can find no examples whatever in the past one hundred years.* There are of course many cases where theories have been found to have a narrower realm of application than had been thought. Newton's theory of motion does not apply at high speeds. Parity, the symmetry between right and left, does not work in the weak forces. And so on. But in this century no theory that has been generally accepted as valid by the world of physics has turned out simply to be a *mistake,* the way that Ptolemy's epicycle theory of planetary motion or the theory that heat is a fluid called caloric were mistakes. Yet in this century, as we have seen in the cases of general relativity and the electroweak theory, the consensus in favor of physical theories has often been reached on the basis of aesthetic judgments before the experimental evidence for these theories became really compelling. I see in this the remarkable power of the physicist's sense of beauty acting in conjunction with and sometimes even in opposition to the weight of experimental evidence.

As I have been describing it, the progress of scientific discovery and validation may seem like pretty much of a muddle. In this respect, there is a nice parallel between the history of war and the history of science. In both cases commentators have searched for systematic rules about how to maximize one's chance of success—that is, for a science of war or a science of science. This may be because in both scientific history and military history, to a far greater degree than in political or cultural or economic history, there is a pretty clear line that one can draw between victory and defeat. We can argue endlessly about the causes and the effects of the American Civil War, but there is no doubt at all that Meade's army beat Lee's at Gettysburg. In the same way, there is no doubt that Copernicus's view of the solar system is better than Ptolemy's, and Darwin's view of evolution is better than Lamarck's.

Even where they do not attempt to formulate a science of

war, military historians often write as if generals lose battles because they do not follow some well-established rules of military science. For instance, two generals of the Union Army in the Civil War that come in for pretty wide disparagement are George McClellan and Ambrose Burnside. McClellan is generally blamed for not being willing to come to grips with the enemy, Lee's Army of Northern Virginia. Burnside is blamed for squandering the lives of his troops in a headlong assault on a well-entrenched opponent at Fredericksburg. It will not escape your attention that McClellan is criticized for not acting like Burnside, and Burnside is criticized for not acting like McClellan. Both Burnside and McClellan were deeply flawed generals, but not because they failed to obey established rules of military science.

The best military historians in fact do recognize the difficulty in stating rules of generalship. They do not speak of a science of war, but rather of a pattern of military behavior that cannot be taught or stated precisely but that somehow or other sometimes helps in winning battles. This is called the art of war. In the same spirit I think that one should not hope for a science of science, the formulation of any definite rules about how scientists do or ought to behave, but only aim at a description of the sort of behavior that historically has led to scientific progress—an art of science.

CHAPTER VI

BEAUTIFUL
THEORIES

When on some gilded cloud or flowre
My gazing soul would dwell an houre,
And in those weaker glories spy
Some shadows of eternity.

Henry Vaughn, *The Retreate*

In 1974 Paul Dirac came to Harvard to speak about his historic work as one of the founders of modern quantum electrodynamics. Toward the end of his talk he addressed himself to our graduate students and advised them to be concerned only with the beauty of their equations, not with what the equations mean. It was not good advice for students, but the search for beauty in physics was a theme that ran throughout Dirac's work and indeed through much of the history of physics.

Some of the talk about the importance of beauty in science has been little more than gushing. I do not propose to use this chapter just to say more nice things about beauty. Rather, I want to focus more closely on the nature of beauty in physical theories, on why our sense of beauty is sometimes a useful guide and sometimes not, and on how the usefulness of our sense of beauty is a sign of our progress toward a final theory.

A physicist who says that a theory is beautiful does not mean quite the same thing that would be meant in saying that a particular painting or a piece of music or poetry is beautiful. It is not merely a personal expression of aesthetic pleasure; it is much closer to what a horse trainer means when he looks at a racehorse and says that it is a beautiful horse. The horse trainer is of course expressing a personal opinion, but it is an opinion about an objective fact: that, on the basis of judgments that the trainer could not easily put into words, this is the kind of horse that wins races.

Of course, different horse trainers may judge a horse differently. That is what makes horse racing. But the horse trainer's aesthetic sense is a means to an objective end—the end of selecting horses that win races. The physicist's sense of beauty is also supposed to serve a purpose—it is supposed to help the physicist select ideas that help us to explain nature. Physicists, just as horse trainers, may be right or wrong in their judgments, but they are not merely enjoying themselves. They often *are* enjoying themselves, but that is not the whole purpose of their aesthetic judgments.

This comparison raises more questions than it answers. First, what *is* a beautiful theory? What are the characteristics of physical theories that give us a sense of beauty? A more difficult question: Why does the physicist's sense of beauty work, when it does work? The stories told in the previous chapter illustrated the rather spooky fact that something as personal and subjective as our sense of beauty helps us not only to invent physical theories but even to judge the validity of theories. Why are we blessed with such aesthetic insight? The effort to answer the question raises another question that is even more difficult, although perhaps it sounds trivial: What is it that the physicist wants to accomplish?

What is a beautiful theory? The curator of a large American art museum once became indignant at my use of the word "beauty" in connection with physics. He said that in his line of work professionals have stopped using this word because they realize how impossible it is to define. Long ago the physicist and mathematician Henri Poincaré admitted that "it may be very hard to define mathematical beauty, but that is just as true of beauty of all kinds."

I will not try to define beauty, any more than I would try to define love or fear. You do not define these things; you know them when you feel them. Later, after the fact, you may sometimes be able to say a little to describe them, as I will try to do here.

By the beauty of a physical theory, I certainly do not mean merely the mechanical beauty of its symbols on the printed page. The metaphysical poet Thomas Traherne took pains that his poems should make pretty patterns on the page, but this is no part of the business of physics. I should also distinguish the sort of beauty I am talking about here from the quality that mathematicians and physicists sometimes call elegance. An elegant proof or calculation is one that achieves a powerful result with a minimum of irrelevant complication. It is not important for the beauty of a theory that its equations should have elegant solutions. The equations of general relativity are notoriously difficult to solve except in the simplest situations, but this does not detract from the beauty of the theory itself. Einstein has been quoted as saying that scientists should leave elegance to tailors.

Simplicity is part of what I mean by beauty, but it is a simplicity of ideas, not simplicity of a mechanical sort that can be measured by counting equations or symbols. Both Einstein's and Newton's theories of gravitation involve equations that tell us the gravitational forces produced by any given amount of matter. In Newton's theory there are three of these equations (corresponding to the three dimensions of space)—in Einstein's theory there are fourteen. In itself, this cannot be counted as an aesthetic advantage of Newton's theory over Einstein's. And in fact it is Einstein's theory that is more beautiful, in part because of the simplicity of his central idea about the equivalence of gravitation and inertia. That is a judgment on which scientists have generally agreed, and as we have seen it was largely responsible for the early acceptance of Einstein's theory.

There is another quality besides simplicity that can make a physical theory beautiful—it is the sense of inevitability that the theory may give us. In listening to a piece of music or hearing a sonnet one sometimes feels an intense aesthetic pleasure at the sense that nothing in the work could be changed, that there is not one note or one word that you would want to have different. In Raphael's *Holy Family* the placement of every figure on the canvas is perfect. This may not be of all paintings in the world your favorite, but as you look at that painting, there is nothing that you would want Raphael to have done differently. The same is partly true (it is never more than partly true) of general relativity. Once you know the general physical principles adopted by Einstein, you understand that there is no other significantly different theory of gravitation to which Einstein could have been led. As Einstein said of general relativity, "The chief attraction of the theory lies in its logical completeness. If a single one of the conclusions drawn from it proves wrong, it must be given up; to modify it without destroying the whole structure seems to be impossible."

This is less true of Newton's theory of gravitation. Newton could have supposed that the gravitational force decreases with the inverse cube of distance rather than the inverse square if that is what the astronomical data had demanded, but Einstein could not have incorporated an inverse-cube law in his theory without scrapping its conceptual basis. Thus Einstein's fourteen equations have an inevitability and hence beauty that Newton's three equations lack. I think that this is what Einstein meant when he referred to the side of the equations that involve the gravitational field in his general theory of relativity as beautiful, as if made of marble, in contrast with the other side of the equations, referring to matter, which he said were still ugly, as if made of mere wood. The way that the gravitational field enters Einstein's equations is almost inevitable, but nothing in general relativity explained why matter takes the form it does.

The same sense of inevitability can be found (again, only in part) in our modern standard model of the strong and electroweak forces that act on elementary particles. There is one common feature that gives both general relativity and the standard model most of their sense of inevitability and simplicity: they obey *principles of symmetry.*

A symmetry principle is simply a statement that something looks the same from certain different points of view. Of all such symmetries, the simplest is the approximate bilateral symmetry of the human face. Because there is little difference between the two sides of your face, it looks the same whether viewed directly or when left and right are reversed, as when you look in a mirror. It is almost a cliché of filmmaking to let the audience realize suddenly that the actor's face they have been watching has been seen in a mirror; the surprise would be spoiled if people had two eyes on the same side of the face like flounders, and always on the same side.

Some things have more extensive symmetries than the human face. A cube looks the same when viewed from six different directions, all at right angles to each other, as well as when left and right are reversed. Perfect crystals look the same not only when viewed from various different directions but also when we shift our positions within the crystal by certain amounts in various directions. A sphere looks the same from any direction. Empty space looks the same from all directions and all positions.

Symmetries like these have amused and intrigued artists and scientists for millennia but did not really play a central role in science. We know many things about salt, and the fact that it is a cubic crystal and therefore looks the same from six different points of view does not rank among the most important. Certainly bilateral symmetry is not the most interesting thing about a human face. The symmetries that are really important in nature are not the symmetries of *things,* but the symmetries of *laws.*

A symmetry of the laws of nature is a statement that when we make certain changes in the point of view from which we observe natural phenomena, the laws of nature we discover do not change. Such symmetries are often called principles of *invariance.* For instance, the laws of nature that we discover take the same form however our laboratories are oriented; it makes no difference whether we measure directions relative to north or northeast or upward or any other direction. This was not so obvious to ancient or medieval natural philosophers; in everyday life there certainly seems to be a difference between up and down and horizontal directions. Only with the birth of modern science in the seventeenth century did it become clear that down seems different from up or north only because below us there happens to be a large mass, the earth, and not (as Aristotle thought) because the natural place of heavy or light things is downward or upward. Note that this symmetry does not say that up is the same as down; observers who measure distances upward or downward from the earth's surface report different descriptions of events such as the fall of an apple, but they discover the same laws, such as the law that apples are attracted by large masses like the earth.

The laws of nature also take the same form wherever our laboratories are located; it makes no difference to our results whether we do our experiments in Texas or Switzerland or on some planet on the other side of the galaxy. The laws of nature take the same form however we set our clocks; it makes no difference whether we date events from the Hegira or the birth of Christ or the beginning of the universe. This does not mean that nothing changes with time or that Texas is just the same as Switzerland, only that the laws discovered at different times and in different places are the same. If it were not for these symmetries the work of science would have to be redone in every new laboratory and in every passing moment.

Any symmetry principle is at the same time a principle of simplicity. If the laws of nature did distinguish among directions like up or down or north, then we would have to put something into our equations to keep track of the orientation of our laboratories, and they would be correspondingly less simple. Indeed, the very notation that is used by mathematicians and physicists to make our equations look as simple and compact as possible has built into it an assumption that all directions in space are equivalent.

Important as these symmetries of the laws of nature are in classical physics, their importance is even greater in quantum mechanics. Consider, what makes one electron different from another? Only its energy, its momentum, and its spin; aside from these properties, every electron in the universe is the same as every other. All these properties of an electron are simply quantities that characterize the way that the quantum-mechanical wave function of the electron responds to symmetry transformations: to changing the way we set our clocks or the location or orientation of our laboratory.* Matter thus loses its

---

* E.g., the frequency with which the wave function of any system in a state of definite energy oscillates is given by the energy divided by a constant of nature known as Planck's constant. This system appears much the same to two ob-

central role in physics: all that is left is principles of symmetry and various ways that wave functions can behave under symmetry transformations.

There are symmetries of space-time that are less obvious than these simple translations or rotations. The laws of nature also appear to take the same form to observers moving at different constant velocities; it makes no difference whether we do our experiments here in the solar system, whizzing around the

---

servers who have set their watches differently by one second, but, if they both observe the system when the hands on their watches both point precisely to twelve noon, they observe that the oscillation is at a different phase; because their watches are set differently they are really observing the system at different times, so that one observer may, e.g., see a crest in the wave, while the other sees a trough. Specifically, the phase is different by the number of cycles (or parts of cycles) that occur in one second; in other words, by the frequency of the oscillation in cycles per second, and hence by the energy divided by Planck's constant. In today's quantum mechanics, we *define* the energy of any system as the change in phase (in cycles or parts of cycles) of the wave function of the system at a given *clock* time when we shift the way our clocks are set by one second. Planck's constant gets into the act only because energy is historically measured in units like calories or kilowatt hours or electron volts that were adopted before the advent of quantum mechanics; Planck's constant simply provides the conversion factor between these older systems of units and the natural quantum-mechanical unit of energy, which is cycles per second. It can be shown that energy defined in this way has all the properties that we normally associate with energy, including its conservation; indeed, the invariance of the laws of nature under the symmetry transformation of resetting our watches is *why* there is such a thing as energy. In much the same way, the component of the momentum of any system in any particular direction is defined as the change of phase of the wave function when we shift the point from which positions are measured by one centimeter in that direction, again times Planck's constant. The amount of spin of a system around any axis is defined as the change of the phase of the wave function when we rotate the frame of reference we use for measuring directions around that axis by one full turn, times Planck's constant. From this point of view, momentum and spin are what they are because of the symmetry of the laws of nature under changes in the frame of reference that we use to measure positions or directions in space. (In listing the properties of electrons I do not include position, because position and momentum are complementary properties; we can describe the state of an electron in terms of its position *or* its momentum but not both together.)

center of the galaxy at hundreds of kilometers per second, or in a distant galaxy rushing away from our own at tens of thousands of kilometers per second. This last symmetry principle is sometimes called the principle of relativity. There is a widespread impression that this principle was invented by Einstein, but there was also a principle of relativity in Newton's theory of mechanics; the difference is only in the way that the speed of the observer affects observations of positions and times in the two theories. But Newton took his version of the principle of relativity for granted; Einstein explicitly designed his version of the principle of relativity to be consistent with an experimental fact, that the speed of light seems the same however the observer is moving. In this sense the emphasis on symmetry as a question of physics in Einstein's 1905 paper on special relativity marks the beginning of the modern attitude to symmetry principles.

The most important difference between the way that observations of space-time positions are affected by the motion of observers in Newton's and Einstein's physics is that in special relativity there is no absolute meaning to a statement that two distant events are simultaneous. One observer may see that two clocks strike noon at the same moment; another observer that is moving with respect to the first finds that one clock strikes noon before or after the other. As pointed out earlier, this makes Newton's theory of gravitation or any similar theory of force inconsistent with special relativity. Newton's theory tells us that the gravitational force that the sun exerts on the earth at any one moment depends on where the sun's mass is at the same moment, but the same moment according to whom?

The natural way to avoid this problem is to abandon the old Newtonian idea of instantaneous action at a distance and to replace it with a picture of force as due to *fields*. In this picture the sun does not directly attract the earth; rather, it creates a field, called the gravitational field, that then exerts a force on the earth. This might seem like a distinction without a difference, but there is a crucial difference: when a solar flare erupts on the sun, it first affects the gravitational field only near the sun, after which the tiny change in the field propagates through space at the speed of light like ripples spreading out from where a pebble falls into water, only reaching the earth some eight minutes later. All observers moving at any constant velocity agree with this description, because in special relativity all such observers agree about the speed of light. In the same way, an electrically charged body creates a field, called the electromagnetic field, that exerts electric and magnetic forces on other charged bodies. When an electrically charged body is suddenly moved, the electromagnetic field is at first changed only near that body, and the changes in this field then propagate at the speed of light. In fact, in this case the changes in the electromagnetic field *are* what we know as light, though it is often light whose wavelength is so low or high that it is not visible to us.

In the context of prequantum physics Einstein's special theory of relativity fit in well with a dualistic view of nature: there are particles, like the electrons, protons, and neutrons in ordinary atoms, and there are fields, like the gravitational or the electromagnetic field. The advent of quantum mechanics led to a much more unified view. Quantum mechanically, the energy and momentum of a field like the electromagnetic field comes in bundles known as photons that behave exactly like particles, though like particles that happen to have no mass. Similarly, the energy and momentum in the gravitational field come in bundles called gravitons, that also behave like particles of zero mass. In a large-scale field of force like the sun's gravitational field we do not notice individual gravitons, essentially because there are so many of them.

In 1929 Werner Heisenberg and Wolfgang Pauli (building

on earlier work of Max Born, Heisenberg, Pascual Jordan, and Eugene Wigner) explained in a pair of papers how massive particles like the electron could also be understood as bundles of energy and momenta in different sorts of fields, such as the electron field. Just as the electromagnetic force between two electrons is due in quantum mechanics to the exchange of photons, the force between photons and electrons is due to the exchange of electrons. The distinction between matter and force largely disappears; any particle can play the role of a test body on which forces act and by its exchange can produce other forces. Today it is generally accepted that the only way to combine the principles of special relativity and quantum mechanics is through the quantum theory of fields or something very like it. This is precisely the sort of logical rigidity that gives a really fundamental theory its beauty: quantum mechanics and special relativity are nearly incompatible, and their reconciliation in quantum field theory imposes powerful restrictions on the ways that particles can interact with one another.

All the symmetries mentioned so far only limit the kinds of force and matter that a theory may contain—they do not in themselves *require* the existence of any particular type of matter or force. Symmetry principles have moved to a new level of importance in this century and especially in the last few decades: there are symmetry principles that dictate the very existence of all the known forces of nature.

In general relativity the underlying principle of symmetry states that *all* frames of reference are equivalent: the laws of nature look the same not only to observers moving at any constant speed but to all observers, whatever the acceleration or rotation of their laboratories. Suppose we move our physical apparatus from the quiet of a university laboratory, and do our experiments on a steadily rotating merry-go-round. Instead of measuring directions relative to north, we would measure them with respect to the horses fixed to the rotating platform. At first sight the laws of nature will appear quite different. Observers on a rotating merry-go-round observe a centrifugal force that seems to pull loose objects to the outside of the merry-go-round. If they are born and grow up on the merry-go-round and do not know that they are on a rotating platform, they describe nature in terms of laws of mechanics that incorporate this centrifugal force, laws that appear quite different from those discovered by the rest of us.

The fact that the laws of nature seem to distinguish between stationary and rotating frames of reference bothered Isaac Newton and continued to trouble physicists in the following centuries. In the 1880s the Viennese physicist and philosopher Ernst Mach pointed the way toward a possible reinterpretation. Mach emphasized that there was something else besides centrifugal force that distinguishes the rotating merry-go-round and more conventional laboratories. From the point of view of an astronomer on the merry-go-round, the sun, stars, galaxies—indeed, the bulk of the matter of the universe—seems to be revolving around the zenith. You or I would say that this is because the merry-go-round is rotating, but an astronomer who grew up on the merry-go-round and naturally uses *it* as his frame of reference would insist that it is the rest of the universe that is spinning around him. Mach asked whether there was any way that this great apparent circulation of matter could be held responsible for centrifugal force. If so, then the laws of nature discovered on the merry-go-round might actually be the same as those found in more conventional laboratories; the apparent difference would simply arise from the different environment seen by observers in their different laboratories.

Mach's hint was picked up by Einstein and made concrete in his general theory of relativity. In general relativity there is indeed an influence exerted by the distant stars that creates the phenomenon of centrifugal force in a spinning merry-go-round: it is the force of gravity. Of course nothing like this happens in Newton's theory of gravitation, which deals only with a simple attraction between all masses. General relativity is more complicated; the circulation of the matter of the universe around the zenith seen by observers on the merry-go-round produces a field somewhat like the magnetic field produced by the circulation of electricity in the coils of an electromagnet. It is this "gravitomagnetic" field that in the merry-go-round frame of reference produces the effects that in more conventional frames of reference are attributed to centrifugal force. The equations of general relativity, unlike those of Newtonian mechanics, are precisely the same in the merry-go-round laboratory and conventional laboratories; the difference between what is observed in these laboratories is entirely due to their different environment—a universe that revolves around the zenith or one that does not. But, if gravitation did not exist, this reinterpretation of centrifugal force would be impossible, and the centrifugal force felt on a merry-go-round would allow us to distinguish between the merry-go-round and more conventional laboratories and would thus rule out any possible equivalence between laboratories that are rotating and those that are not. *Thus the symmetry among different frames of reference requires the existence of gravitation.*

The symmetry underlying the electroweak theory is a little more esoteric. It does not have to do with changes in our point of view in space and time but rather with changes in our point of view about the identity of the different types of elementary particle. Just as it is possible for a particle to be in a quantum-mechanical state in which it is neither definitely here nor there, or spinning neither definitely clockwise nor counterclockwise, so also through the wonders of quantum mechanics it is possible to have a particle in a state in which it is neither definitely an electron nor definitely a neutrino until we measure some property that would distinguish the two, like the electric charge. In the electroweak theory the form of the laws of nature is unchanged if we replace electrons and neutrinos everywhere in our equations with such mixed states that are neither electrons nor neutrinos. Because various other types of particles interact with electrons and neutrinos, it is necessary at the same time to mix up families of these other particle types, such as up quarks with down quarks, as well as the photons with its siblings, the positively and negatively charged W particles and the neutral Z particles. This is the symmetry that connects the electromagnetic forces, which are produced by an exchange of photons, with the weak nuclear forces that are produced by the exchange of the W and Z particles. The photon and the W and Z particles appear in the electroweak theory as bundles of the energy of four fields, fields that are required by this symmetry of the electroweak theory in much the same way that the gravitational field is required by the symmetries of general relativity.

Symmetries of the sort underlying the electroweak theory are called *internal symmetries,* because we can think of them as having to do with the intrinsic nature of the particles, rather than their position or motion. Internal symmetries are less familiar than those that act on ordinary space and time, such as those governing general relativity. You can think of each particle as carrying a little dial, with a pointer that points in directions marked "electron" or "neutrino" or "photon" or "W" or anywhere in between. The internal symmetry says that the laws of nature take the same form if we rotate the markings on these dials in certain ways.

Furthermore, for the sort of symmetry that governs the electroweak forces, we can rotate the dials differently for particles at different times or positions. This is much like the symmetry underlying general relativity, which allows us to rotate our laboratory not only by some fixed angle but also by an amount that increases with time, by placing it on a merry-go-round. The invariance of the laws of nature under a group of such position-dependent and time-dependent internal symmetry transformations is called a *local* symmetry (because the effect of the symmetry transformations depends on location in space and time) or a *gauge* symmetry (for reasons that are purely historical). It is the local symmetry between different frames of reference in space and time that makes gravitation necessary, and in much the same way it is a second local symmetry between electrons and neutrinos (and between up quarks and down quarks and so on) that makes the existence of the photon, W, and Z fields necessary.

There is yet a third exact local symmetry associated with the internal property of quarks that is fancifully known as *color*. We have seen that there are quarks of various types, like the up and down quarks that make up the protons and neutrons that are found in all ordinary atomic nuclei. In addition, each of these types of quarks comes in three different "colors," which physicists (at least in the United States) often call red, white, and blue. Of course, this has nothing to do with ordinary color; it is merely a label used to distinguish different subvarieties of quark. As far as we know, there is an exact symmetry in nature among the different colors; the force between a red quark and a white quark is the same as between a white quark and a blue quark, and the force between two red quarks is the same as between two blue quarks. But this symmetry goes beyond mere interchanges of colors. In quantum mechanics we can consider states of a single quark that are neither definitely red nor definitely white nor definitely blue. The laws of nature take precisely the same form if we replace red, white, and blue quarks with quarks in three suitable mixed states (e.g., purple, pink, and lavender). Again in analogy with general relativity, the fact that the laws of nature are unaffected even if the mixtures vary from place to place and time to time makes it necessary to include a family of fields in the theory that interact with quarks, analogous to the gravitational field. There are eight of these fields; they are known as gluon fields because the strong forces they produce glue the quarks together inside the proton and neutron. Our modern theory of these forces, *quantum chromodynamics,* is nothing but the theory of quarks and gluons that respects this local color symmetry. The standard model of elementary particles consists of the electroweak theory combined with quantum chromodynamics.

I have been referring to principles of symmetry as giving theories a kind of rigidity. You might think that this is a drawback, that the physicist wants to develop theories that are capable of describing a wide variety of phenomena and therefore would like to discover theories that are as flexible as possible—theories that make sense under a wide variety of possible circumstances. That is true in many areas of science, but it is not true in this kind of fundamental physics. We are on the track of something universal—something that governs physical phenomena throughout the universe—something that we call the laws of nature. We do not want to discover a theory that is capable of describing all imaginable kinds of force among the particles of nature. Rather, we hope for a theory that rigidly will allow us to describe only those forces—gravitational, electroweak, and strong—that actually as it happens do exist. This kind of rigidity in our physical theories is part of what we recognize as beauty.

It is not only principles of symmetry that give rigidity to our theories. On the basis of symmetry principles alone we would not be led to the electroweak theory or quantum chromodynamics, except as a special case of a much larger variety of theories with a unlimited number of adjustable constants that could be put into the theory with any values we like. The additional constraint that allowed us to pick out our simple standard model out of the variety of other more complicated theories that satisfy the same symmetry principles was the condition that infinities that arise in calculations using the theory should all cancel. (That is, the theory must be "renormalizable.") This condition turns out to impose a high degree of simplicity on the equations of the theory and, together with the various local symmetries, goes a long way to giving a unique shape to our standard model of elementary particles.

The beauty that we find in physical theories like general relativity or the standard model is very like the beauty conferred on some works of art by the sense of inevitability that they give us—the sense that one would not want to change a note or a brush stroke or a line. But just as in our appreciation of music or painting or poetry, this sense of inevitability is a matter of taste and experience and cannot be reduced to formula.

Every other year the Lawrence Berkeley Laboratory publishes a little booklet that lists the properties of the elementary particles as known to that date. If I say that the fundamental principle governing nature is that the elementary particles have the properties listed in this booklet, then it is certainly true that the known properties of the elementary particles follow inevitably from this fundamental principle. This principle even has predictive power—every new electron or proton created in our laboratories will be found to have the mass and charge already listed in the booklet. But the principle itself is so ugly that no one would feel that anything had been accomplished. Its ugliness lies in its lack of simplicity and inevitability—the booklet contains thousands of numbers, any one of which could be changed without making nonsense of the rest of the information. There is no logical formula that establishes a sharp dividing line between a beautiful explanatory theory and a mere list of data, but we know the difference when we see it—we demand a simplicity and rigidity in our principles before we are willing to take them seriously. Thus not only is our aesthetic judgment a means to the end of finding scientific explanations and judging their validity—*it is part of what we mean by an explanation.*

Other scientists sometimes poke fun at elementary particle physicists because there are now so many so-called elementary particles that we have to carry the Berkeley booklet around with us to remind us of all the particles that have been discovered. But the mere number of particles is not important. As Abdus Salam has said, it is not particles or forces with which nature is sparing, but principles. The important thing is to have a set of simple and economical principles that explain why the particles are what they are. It *is* disturbing that we do not yet have a complete theory of the sort we want. But, when we do, it will not matter very much how many kinds of particle or force it describes, as long as it does so beautifully, as an inevitable consequence of simple principles.

The kind of beauty that we find in physical theories is of a very limited sort. It is, as far as I have been able to capture it in words, the beauty of simplicity and inevitability—the beauty of perfect structure, the beauty of everything fitting together, of nothing being changeable, of logical rigidity. It is a beauty that is spare and classic, the sort we find in the Greek tragedies. But this is not the only kind of beauty that we find in the arts. A play of Shakespeare does not have this beauty, at any rate not to the extent that some of his sonnets have. Often the director of a Shakespeare play chooses to leave out whole speeches. In the Olivier film version of *Hamlet,* Hamlet never says, "Oh what a rogue and peasant slave am I! . . ." And yet the performance works, because Shakespeare's plays are not spare, perfect structures like general relativity or *Oedipus Rex;* they are big messy compositions whose messiness mirrors the complexity of life. That is part of the beauty of his plays, a beauty that to my taste is of a higher order than the beauty of a play of Sophocles or the beauty of general relativity for that matter. Some of the greatest moments in Shakespeare are those in which he deliberately abandons the model of Greek tragedy and introduces an extraneous comic proletarian—a doorkeeper or gardener or fig seller or grave digger—just before his major characters encounter their fates. Certainly the beauty of theoretical physics would be a very bad exemplar for the arts, but such as it is it gives us pleasure and guidance.

There is another respect in which it seems to me that theoretical physics is a bad model for the arts. Our theories are very esoteric—necessarily so, because we are forced to develop these theories using a language, the language of mathematics, that has not become part of the general equipment of the educated public. Physicists generally do not like the fact that our theories are so esoteric. On the other hand, I have occasionally heard artists talk proudly about their work being accessible only to a band of cognoscenti and justify this attitude by quoting the example of physical theories like general relativity that also can be understood only by initiates. Artists like physicists may not always be able to make themselves understood by the general public, but esotericism for its own sake is just silly.

Although we seek theories that are beautiful because of a rigidity imposed on them by simple underlying principles, creating a theory is not simply a matter of deducing it mathematically from a set of preordained principles. Our principles are often invented as we go along, sometimes precisely because they lead to the kind of rigidity we hope for. I have no doubt that one of the reasons that Einstein was so pleased with his idea about the equivalence of gravitation and inertia was that this principle led to only one fairly rigid theory of gravitation and not to an infinite variety of possible theories of gravitation. Deducing the consequences of a given set of well-formulated physical principles can be difficult or easy, but it is the sort of thing that physicists learn to do in graduate school and that they generally enjoy doing. The creation of *new* physical principles is agony and apparently cannot be taught.

Weirdly, although the beauty of physical theories is embodied in rigid mathematical structures based on simple underlying principles, the structures that have this sort of beauty tend to survive even when the underlying principles are found to be wrong. A good example is Dirac's theory of the electron. Dirac in 1928 was trying to rework Schrödinger's version of quantum mechanics in terms of particle waves so that it would be consistent with the special theory of relativity. This effort led Dirac to the conclusions that the electron must have a certain spin, and that the universe is filled with unobservable electrons of negative energy, whose *absence* at a particular point would be seen in the laboratory as the presence of an electron with the opposite charge, that is, an antiparticle of the electron. His theory gained an enormous prestige from the 1932 discovery in cosmic rays of precisely such an antiparticle of the electron, the particle now called the positron. Dirac's theory was a key ingredient in the version of quantum electrodynamics that was developed and applied with great success in the 1930s and 1940s. But we know today that Dirac's point of view was largely wrong. The proper context for the reconciliation of quantum mechanics and special relativity is not the sort of relativistic version of Schrödinger's wave mechanics that Dirac sought, but the more general formalism known as quantum field theory, presented by Heisenberg and Pauli in 1929. In quantum field theory not only is the photon a bundle of the energy of a field, the electromagnetic field; so also the electron and positrons are bundles of the energy of the electron field, and all other elementary particles are bundles of the energy of various other fields. Almost by accident, Dirac's theory of the electron gave the same results as quantum field theory for processes involving only electrons, positrons, and/or photons. But quantum field theory is more general—it can account for processes like nuclear beta decay that could not be understood along the lines of Dirac's theory. There is nothing in quantum field theory that requires particles to have any particular spin. The electron does happen to have the spin that Dirac's theory required, but there are other particles with other spins and those other particles have antiparticles and this has nothing to do with the negative energies about which Dirac speculated. Yet the *mathematics* of Dirac's theory has survived as an essential part of quantum field theory; it must be taught in every graduate course in advanced quantum mechanics. The formal structure of Dirac's theory has thus survived the death of the principles of relativistic wave mechanics that Dirac followed in being led to his theory.

So the mathematical structures that physicists develop in obedience to physical principles have an odd kind of portability. They can be carried over from one conceptual environment to another and serve many different purposes, like the clever bones in your shoulders that in another animal would be the joint between the wing and the body of a bird or the flipper and body of a dolphin. We are led to these beautiful structures by physical principles, but the beauty sometimes survives when the principles themselves do not.

A possible explanation was given by Niels Bohr. In speculating in 1922 about the future of his earlier theory of atomic structure, he remarked that "mathematics has only a limited number of forms which we can adapt to Nature, and it can happen to one that he finds the right forms by formulating entirely wrong concepts." As it happened, Bohr was right about the future of his own theory; its underlying principles have been abandoned, but we still use some of its language and methods of calculation.

It is precisely in the application of pure mathematics to physics that the effectiveness of aesthetic judgments is most amazing. It has become a commonplace that mathematicians are driven in their work by the wish to construct formalisms that are conceptually beautiful. The English mathematician G. H. Hardy explained that "mathematical patterns like those of the painters or the poets must be beautiful. The ideas, like the colors or the words must fit together in a harmonious way. Beauty is the first test. There is no permanent place for ugly mathematics." And yet mathematical structures that confessedly are developed by mathematicians because they seek a sort of beauty are often found later to be extraordinarily valuable by the physicist.

For illustration, let us return to the example of non-Euclidean geometry and general relativity. After Euclid, mathematicians tried for two millennia to learn whether the different assumptions underlying Euclid's geometry were logically independent of each other. If the postulates were not independent, if some of them could be deduced from the others, then the unnecessary postulates could be dropped, yielding a more economical and hence more beautiful formulation of geometry. This effort came to a head in the early years of the nineteenth century, when "the prince of geometers" Carl Friedrich Gauss and others developed a non-Euclidean geometry for a sort of curved space that satisfied all Euclid's postulates except the fifth. This showed that Euclid's fifth postulate is indeed logically independent of the other postulates. The new geometry was developed in order to settle a historic question about the foundations of geometry, not at all because anyone thought it applied to the real world.

Non-Euclidean geometry was then extended by one of the

greatest of all mathematicians, Georg Friedrich Bernhard Riemann, to a general theory of curved spaces of two or three or any number of dimensions. Mathematicians continued to work on Riemannian geometry because it was so beautiful, without any idea of physical applications. Its beauty was again largely the beauty of inevitability. Once you start thinking about curved spaces, you are led almost inevitably to the introduction of the mathematical quantities ("metrics," "affine connections," "curvature tensors," and so on) that are the ingredients of Riemannian geometry. When Einstein started to develop general relativity, he realized that one way of expressing his ideas about the symmetry that relates different frames of reference was to ascribe gravitation to the curvature of space-time. He asked a friend, Marcel Grossman, whether there existed any mathematical theory of curved spaces—not merely of curved two-dimensional surfaces in ordinary Euclidean three-dimensional space, but of curved three-dimensional spaces, or even curved four-dimensional space-times. Grossman gave Einstein the good news that there did in fact exist such a mathematical formalism, the one developed by Riemann and others, and taught him this mathematics, which Einstein then incorporated into general relativity. The mathematics was there waiting for Einstein to make use of, although I believe that Gauss and Riemann and the other differential geometers of the nineteenth century had no idea that their work would ever have any application to physical theories of gravitation.

An even stranger example is provided by the history of internal symmetry principles. In physics internal symmetry principles typically impose a kind of family structure on the menu of possible particles. The first known example of such a family was provided by the two types of particle that make up ordinary atomic nuclei, the proton and neutron. Protons and neutrons have very nearly the same mass, so, once the neutron was discovered by James Chadwick in 1932, it was natural to suppose that the strong nuclear forces (which contribute to the neutron and proton masses) should respect a simple symmetry: the equations governing these forces should preserve their form if everywhere in these equations the roles of neutrons and protons are reversed. This would tell us among other things that the strong nuclear force is the same between two neutrons as between two protons but would tell us nothing about the force between a proton and a neutron. It was therefore somewhat a surprise when experiments in 1936 revealed that the nuclear force between two protons is about the same as the force between a proton and a neutron. This observation gave rise to the idea of a symmetry that goes beyond mere interchanges of protons and neutrons, a symmetry under continuous transformations that change protons and neutrons into particles that are proton-neutron mixtures, with arbitrary probabilities of being a proton or a neutron.

These symmetry transformations act on the particle label that distinguishes protons and neutrons in a way that is mathematically the same as the way that ordinary rotations in three dimensions act on the spins of particles like protons or neutrons or electrons. With this example in mind, until the 1960s many physicists tacitly assumed that the internal symmetry transformations that leave the laws of nature unchanged had to take the form of rotations in some internal space of two, three, or more dimensions, like the rotations of protons and neutrons into one another. The textbooks on the application of symmetry principles to physics then available (including the classic books of Hermann Weyl and Eugene Wigner) barely gave any hint that there were other mathematical possibilities. It was not until a host of new particles was discovered in cosmic rays and then at accelerators like the Bevatron in Berkeley in the late 1950s that a wider view of the possibilities of internal symmetries was forced on the world of theoretical physics. These particles seemed to fall into families that were more extensive than the simple proton-neutron pair of twins. For instance, the neutron and proton were found to bear a strong family likeness to six other particles known as hyperons, of the same spin and similar mass. What sort of internal symmetry could give rise to such extended kinship groups?

Around 1960 physicists studying this question began to turn for help to the literature of mathematics. It came to them as a delightful surprise that mathematicians had in a sense already cataloged all possible symmetries. The complete set of transformations that leaves anything unchanged, whether a specific object or the laws of nature, forms a mathematical structure known as a *group,* and the general mathematics of symmetry transformations is known as *group theory.* Each group is characterized by abstract mathematical rules that do not depend on what it is that is being transformed, just as the rules of arithmetic do not depend on what it is we are adding or multiplying. The menu of the types of families that are allowed by any particular symmetry of the laws of nature is completely dictated by the mathematical structure of the symmetry group.

Those groups of transformations that act continuously, like rotations in ordinary space or the mixing of electrons and neutrinos in the electroweak theory, are called *Lie groups,* after the Norwegian mathematician Sophus Lie. The French mathematician Élie Cartan had in his 1894 thesis given a list of all the "simple" Lie groups, from which all others could be built up by combining their transformations. In 1960 Gell-Mann and the Israeli physicist Yuval Ne'eman independently found that one of these simple Lie groups (known as SU(3)) was just right to impose a family structure on the crowd of elementary particles much like what had been found experimentally. Gell-Mann borrowed a term from Buddhism and called this symmetry principle the eightfold way, because the better-known particles fell into families with eight members, like the neutron, proton, and their six siblings. Not all families were then complete; a new particle was needed to complete a family of ten particles that are similar to neutrons and protons and hyperons but have three times higher spin. It was one of the great successes of the new SU(3) symmetry that this predicted particle was subsequently discovered in 1964 at Brookhaven and turned out to have the mass estimated by Gell-Mann.

Yet this group theory that turned out to be so relevant to physics had been invented by mathematicians for reasons that were strictly internal to mathematics. Group theory was initiated in the early nineteenth century by Evariste Galois, in his proof that there are no general formulas for the solution of certain algebraic equations (equations that involve fifth or higher powers of the unknown quantity). Neither Galois nor Lie nor Cartan had any idea of the sort of application that group theory would have in physics.

It is very strange that mathematicians are led by their sense of mathematical beauty to develop formal structures that physicists only later find useful, even where the mathematician had no such goal in mind. A well-known essay by the physicist Eugene Wigner refers to this phenomenon as "The Unreasonable Effectiveness of Mathematics." Physicists generally find the ability of mathematicians to anticipate the mathematics needed in the theories of physicists quite uncanny. It is as if Neil Armstrong in 1969 when he first set foot on the surface of the moon had found in the lunar dust the footsteps of Jules Verne.

Where then *does* a physicist get a sense of beauty that helps not only in discovering theories of the real world, but even in judging the validity of physical theories, sometimes in the teeth of contrary experimental evidence? And how does a mathematician's sense of beauty lead to structures that are valuable decades or centuries later to physicists, even though the mathematician may have no interest in physical applications?

There seem to me to be three plausible explanations, two of them applicable throughout much of science and the third limited to the most fundamental areas of physics. The first explanation is that the universe itself acts on us as a random, inefficient, and yet in the long run effective, teaching machine. Just as through an infinite series of accidental events, atoms of carbon and nitrogen and oxygen and hydrogen joined together to form primitive forms of life that later evolved into protozoa and fishes and people, in the same manner our way of looking at the universe has gradually evolved through a natural selection of ideas. Through countless false starts, we have gotten it beaten into us that nature is a certain way, and we have grown to look at that way that nature is as beautiful.

I suppose this would be everyone's explanation of why the horse trainer's sense of beauty helps when it does help in judging which horse can win races. The racehorse trainer has been at the track for many years—has experienced many horses winning or losing—and has come to associate, without being able to express it explicitly, certain visual cues with the expectation of a winning horse.

One of the things that makes the history of science so endlessly fascinating is to follow the slow education of our species in the sort of beauty to expect in nature. I once went back to the original literature of the 1930s on the earliest internal symmetry principle in nuclear physics, the symmetry that I mentioned earlier between neutrons and protons, to try to find the one research article that first presented this symmetry principle the way it would be presented today, as a fundamental fact about nuclear physics that stands on its own, independent of any detailed theory of nuclear forces. I could find no such article. It seems that in the 1930s it was simply not good form to write papers based on symmetry principles. What was good form was to write papers about nuclear forces. If the forces turned out to have a certain symmetry, so much the better, for, if you knew the proton-neutron force, you did not have to guess the proton-proton force. But the symmetry principle itself was not regarded, as far as I can tell, as a feature that would legitimize a theory—that would make the theory beautiful. Symmetry principles were regarded as mathematical tricks; the real business of physicists was to work out the dynamical details of the forces we observe.

We feel differently today. If experimenters were to discover some new particles that formed families of some sort or other like the proton-neutron doublet, then the mail would instantly be filled with hundreds of preprints of theoretical articles speculating about the sort of symmetry that underlies this family structure, and, if a new kind of force were discovered, we would all start speculating about the symmetry that dictates the existence of that force. Evidently we have been changed by the universe acting as a teaching machine and imposing on us a sense of beauty with which our species was not born.

Even mathematicians live in the real universe, and respond to its lessons. Euclid's geometry was taught to schoolchildren for two millennia as a nearly perfect example of abstract deductive reasoning, but we learned in this century from general relativity that Euclidean geometry works as well as it does only because the gravitational field on the surface of the earth is rather weak, so that the space in which we live has no noticeable curvature. In formulating his postulates Euclid was in fact acting as a physicist, using his experience of life in the weak gravitational fields of Hellenistic Alexandria to make a theory of uncurved space. He did not know how limited and contingent his geometry was. Indeed, it is only relatively recently that we have learned to make a distinction between pure mathematics and the science to which it is applied. The Lucasian Chair at Cambridge that was held by Newton and Dirac was (and still is) officially a professorship in mathematics, not physics. It was not until the development of a rigorous and abstract mathematical style by Augustin-Louis Cauchy and others in the early nineteenth century that mathematicians took as an ideal that their work should be independent of experience and common sense.

The second of the reasons why we expect successful scientific theories to be beautiful is simply that scientists tend to choose problems that are likely to have beautiful solutions. The same may even apply to our friend the racehorse trainer. He trains horses to win races; he has learned to recognize which horses are likely to win and he calls these horses beautiful; but, if you take him aside and promise not to repeat what he says, he may confess to you that the reason he went into the business of training horses to win races in the first place was because the horses that he trains are such beautiful animals.

A good example in physics is provided by the phenomenon of smooth phase transitions,* like the spontaneous disappearance of magnetism when an iron permanent magnet is heated above a temperature of 770°C, the temperature known as the Curie point. Because this is a smooth transition, the magnetization of a piece of iron goes to zero gradually as the tempera-

---

*What I call "smooth" phase transitions are also often called "second-order phase transitions." This is to distinguish them from "first-order phase transitions," like the boiling of water at 100°C or the melting of ice at 0°C, in which the properties of the material change discontinuously. It takes a certain amount of energy (the so-called latent heat) to convert ice at 0°C to liquid water at the same temperature, or liquid water at 100°C to water vapor at the same temperature, but it takes no extra energy to wipe out the magnetism of a piece of iron when the temperature is just at the Curie point.

ture approaches the Curie point. The surprising thing about such phase transitions is the *way* that the magnetization goes to zero. Estimates of various energies in a magnet had led physicists to expect that, when the temperature is only slightly below the Curie point, the magnetization would be simply proportional to the square root of the difference between the Curie point and the temperature. Instead it was observed experimentally that the magnetization is proportional to the 0.37 power of this difference. That is, the dependence of the magnetization on the temperature is somewhere between being proportional to the square root (the 0.5 power) and the cube root (the 0.33 power) of the difference between the Curie point and the temperature.

Powers like this 0.37 are called *critical exponents,* sometimes with the adjective "nonclassical" or "anomalous," because they are not what had been expected. Other quantities were observed to behave in similar ways in this and other phase transitions, in some cases with precisely the same critical exponents. This is not an intrinsically glamorous phenomenon, like black holes or the expansion of the universe. Nevertheless, some of the brightest theoretical physicists in the world worked on the problem of the critical exponents, until the problem was eventually solved in 1972 by Kenneth Wilson and Michael Fisher, both then at Cornell. Yet it might have been thought that the precise calculation of the Curie point itself was a problem of greater practical importance. Why should leaders of condensed matter theory give the problem of the critical exponents so much greater priority?

I think that the problem of critical exponents attracted so much attention because physicists judged that it would be likely to have a beautiful solution. The clues that suggested that the solution would be beautiful were above all the universality of the phenomenon, the fact that the same critical exponents would crop up in very different problems, and also the fact that physicists have become used to finding that the most essential properties of physical phenomena are often expressed in terms of laws that relate physical quantities to powers of other quantities, such as the inverse-square law of gravitation. As it turned out, the theory of critical exponents has a simplicity and inevitability that makes it one of the most beautiful in all of physics. In contrast, the problem of calculating the precise temperatures of phase transitions is a messy one, whose solution involves complicated details of the iron or other substance that undergoes the phase transition, and for this reason it is studied either because of its practical importance or in want of anything better to do.

In some cases the initial hopes of scientists for a beautiful theory have turned out to be misplaced. A good example is provided by the genetic code. Francis Crick describes in his autobiography how after the discovery of the double-helix structure of DNA by himself and James Watson, the attention of molecular biologists turned to breaking the code by which the cell interprets the sequence of chemical units on the two helices of DNA as a recipe for building suitable protein molecules. It was known that proteins are built up out of chains of amino acids, that there are only twenty amino acids that are important in virtually all plants and animals, that the information for selecting each successive amino acid in a protein molecule is carried by the choices of three successive pairs of chemical units called bases, of which there are only four different kinds. So the genetic code interprets three successive choices each out of four possible base pairs (like three cards chosen in order from a deck of cards that show only the four suits but no numbers or faces) to dictate each choice of one out of twenty possible amino acids to be added to the protein. Molecular biologists invented all sorts of elegant principles that might govern this code—for instance, that no information in the choice of three base pairs would be wasted, and that any information not needed to specify an amino acid would be used for error detection, like the extra bits that are sent between computers to check the accuracy of the transmission. The answer found in the early 1960s turned out to be very different. The genetic code is pretty much a mess; some amino acids are called for by more than one triplet of base pairs, and some triplets produce nothing at all. The genetic code is not as bad as a randomly chosen code, which suggests that it has been somewhat improved by evolution, but any communications engineer could design a better code. The reason of course is that the genetic code was *not* designed; it developed through a series of accidents at the beginning of life on earth and has been inherited in more or less this form by all subsequent organisms. Of course the genetic code is so important to us that we study it whether it is beautiful or not, but it is a little disappointing that it did not turn out to be beautiful.

Sometimes when our sense of beauty lets us down, it is because we have overestimated the fundamental character of what we are trying to explain. A famous example is the work of the young Johannes Kepler on the sizes of the orbits of planets.

Kepler was aware of one of the most beautiful conclusions of Greek mathematics, concerning what are called the Platonic solids. These are three-dimensional objects with plane boundaries, for which every vertex and every face and every line are exactly like every other vertex, face, and line. An obvious example is the cube. The Greeks discovered that there are all together only five of these Platonic solids: the cube, the triangular pyramid, the twelve-sided dodecahedron, the eight-sided octahedron, and the twenty-sided icosahedron. (They are called Platonic solids because Plato in the *Timaeus* proposed a one-to-one correspondence between them and the supposed five elements, a view subsequently attacked by Aristotle.) The Platonic solids furnish a prime example of mathematical beauty; this discovery has the same sort of beauty as the Cartan catalog of all possible continuous symmetry principles.

Kepler in his *Mysterium cosmographicum* proposed that the existence of only five Platonic solids explained why there were (apart from the earth) just five planets: Mercury, Venus, Mars, Jupiter, and Saturn. (Uranus, Neptune, and Pluto were not discovered until later.) To each one of these five planets Kepler associated one of the Platonic solids, and he made the guess that the radius of each planet's orbit was in proportion to the radius of the corresponding Platonic solid when the solids are nested within each other in the right order. Kepler wrote that he had worked over the irregularities of planetary motion "until they were at last accommodated to the laws of nature."

To a scientist today it may seem a scandal that one of the founders of modern science should invent such a fanciful model of the solar system. This is not only because Kepler's scheme did not fit observations of the solar system (though it did not), but much more because we know that this is not the sort of speculation that is appropriate to the solar system. But Kepler was not a fool. The kind of speculative reasoning he applied to the solar system is very similar to the sort of theorizing that elementary particle physicists do today; we do not associate anything with the Platonic solids, but we do believe for instance in a correspondence between different possible kinds of force and different members of the Cartan catalog of all possible symmetries. Where Kepler went wrong was not in using this sort of guesswork, but in supposing (as most philosophers before him had supposed) that the planets are important.

Of course, the planets are important in some ways. We live on one of them. But their existence is not incorporated into the laws of nature at any fundamental level. We now understand that the planets and their orbits are the results of a sequence of historical accidents and that, although physical theory can tell us which orbits are stable and which would be chaotic, there is no reason to expect any relations among the sizes of their orbits that would be mathematically simple and beautiful.

It is when we study truly fundamental problems that we expect to find beautiful answers. We believe that, if we ask why the world is the way it is and then ask why that answer is the way it is, at the end of this chain of explanations we shall find a few simple principles of compelling beauty. We think this in part because our historical experience teaches us that as we look beneath the surface of things, we find more and more beauty. Plato and the neo-Platonists taught that the beauty we see in nature is a reflection of the beauty of the ultimate, the *nous*. For us, too, the beauty of present theories is an anticipation, a premonition, of the beauty of the final theory. And in any case, we would not accept any theory as final unless it were beautiful.

Although we do not yet have a sure sense of where in our work we should rely on our sense of beauty, still in elementary particle physics aesthetic judgments seem to be working increasingly well. I take this as evidence that we are moving in the right direction, and perhaps not so far from our goal.

CHAPTER VII

AGAINST
PHILOSOPHY

Myself when young did eagerly frequent
Doctor and Saint, and heard great argument
About it and about: but evermore
Came out by the same door wherein I went.

Edward FitzGerald, *Rubáiyát of Omar Khayyám*

Physicists get so much help from subjective and often vague aesthetic judgments that it might be expected that we would be helped also by philosophy, out of which after all our science evolved. Can philosophy give us any guidance toward a final theory?

The value today of philosophy to physics seems to me to be something like the value of early nation-states to their peoples. It is only a small exaggeration to say that, until the introduction of the post office, the chief service of nation-states was to protect their peoples from other nation-states. The insights of philosophers have occasionally benefited physicists, but generally in a negative fashion—by protecting them from the preconceptions of other philosophers.

I do not want to draw the lesson here that physics is best done without preconceptions. At any one moment there are so many things that might be done, so many accepted principles that might be challenged, that without some guidance from our preconceptions one could do nothing at all. It is just that philosophical principles have not generally provided us with the right preconceptions. In our hunt for the final theory, physicists are more like hounds than hawks; we have become good at sniffing around on the ground for traces of the beauty we expect in the laws of nature, but we do not seem to be able to see the path to the truth from the heights of philosophy.

Physicists do of course carry around with them a working philosophy. For most of us, it is a rough-and-ready realism, a belief in the objective reality of the ingredients of our scientific theories. But this has been learned through the experience of scientific research and rarely from the teachings of philosophers.

This is not to deny all value to philosophy, much of which has nothing to do with science. I do not even mean to deny all value to the philosophy of science, which at its best seems to me a pleasing gloss on the history and discoveries of science. But we should not expect it to provide today's scientists with any useful guidance about how to go about their work or about what they are likely to find.

I should acknowledge that this is understood by many of the philosophers themselves. After surveying three decades of professional writings in the philosophy of science, the philosopher George Gale concludes that "these almost arcane discussions, verging on the scholastic, could have interested only the smallest number of practicing scientists." Wittgenstein remarked that "nothing seems to me less likely than that a scientist or mathematician who reads me should be seriously influenced in the way he works."

This is not merely a matter of the scientist's intellectual laziness. It is agonizing to have to interrupt one's work to learn a new discipline, but scientists do it when we have to. At various times I have managed to take time off from what I was doing to learn all sorts of things I needed to know, from differential topology to Microsoft DOS. It is just that a knowledge of philosophy does not seem to be of use to physicists—always with the exception that the work of some philosophers helps us to avoid the errors of other philosophers.

It is only fair to admit my limitations and biases in making this judgment. After a few years' infatuation with philosophy as an undergraduate I became disenchanted. The insights of the philosophers I studied seemed murky and inconsequential compared with the dazzling successes of physics and mathematics. From time to time since then I have tried to read current work on the philosophy of science. Some of it I found to be written in a jargon so impenetrable that I can only think that it aimed at impressing those who confound obscurity with profundity. Some of it was good reading and even witty, like the writings of Wittgenstein and Paul Feyerabend. But only rarely did it seem to me to have anything to do with the work of science as I knew it. According to Feyerabend, the notion of scientific explanation developed by some philosophers of science is so narrow that it is impossible to speak of one theory being explained by another, a view that would leave my generation of particle physicists with nothing to do.

It may seem to the reader (especially if the reader is a professional philosopher) that a scientist who is as out of tune with the philosophy of science as I am should tiptoe gracefully past the subject and leave it to experts. I know how philosophers feel about attempts by scientists at amateur philosophy. But I do not aim here to play the role of a philosopher, but rather that of a specimen, an unregenerate working scientist who finds no help in professional philosophy. I am not alone in this; I know of *no one* who has participated actively in the advance of physics in the postwar period whose research has been significantly helped by the work of philosophers. I raised in the previous chapter the problem of what Wigner calls the "unreasonable effectiveness" of mathematics; here I want to take up another equally puzzling phenomenon, the unreasonable ineffectiveness of philosophy.

Even where philosophical doctrines have in the past been useful to scientists, they have generally lingered on too long, becoming of more harm than ever they were of use. Take, for example, the venerable doctrine of "mechanism," the idea that nature operates through pushes and pulls of material particles or fluids. In the ancient world no doctrine could have been more progressive. Ever since the pre-Socratic philosophers Democritus and Leucippus began to speculate about atoms, the idea that natural phenomena have mechanical causes has stood in opposition to popular beliefs in gods and demons. The Hellenistic cult leader Epicurus brought a mechanical worldview into his creed specifically as an antidote to belief in the Olympian gods. When René Descartes set out in the 1630s on his great attempt to understand the world in rational terms, it was natural that he should describe physical forces like gravitation in a mechanical way, in terms of vortices in a material fluid filling all space. The "mechanical philosophy" of Descartes had a powerful influence on Newton, not because it was right (Descartes did not seem to have the modern idea of testing theories quantitatively) but because it provided an example of the sort of mechanical theory that could make sense out of nature. Mechanism reached its zenith in the nineteenth century, with the brilliant explanation of chemistry and heat in terms of atoms. And even today mechanism seems to many to be simply the logical opposite to superstition. In the history of human thought the mechanical worldview has played a heroic role.

That is just the trouble. In science as in politics or economics we are in great danger from heroic ideas that have outlived their usefulness. The heroic past of mechanism gave it such prestige that the followers of Descartes had trouble accepting Newton's theory of the solar system. How could a good Cartesian, believing that all natural phenomena could be reduced to the impact of material bodies or fluids on one another, accept Newton's view that the sun exerts a force on the earth across 93 million miles of empty space? It was not until well into the eighteenth century that Continental philosophers began to feel comfortable with the idea of action at a distance. In the end Newton's ideas did prevail on the Continent as well as in Britain, in Holland, Italy, France, and Germany (in that order) from 1720 on. To be sure, this was partly due to the influence of philosophers like Voltaire and Kant. But here again the service of philosophy was a negative one; it helped only to free science from the constraints of philosophy itself.

Even after the triumph of Newtonianism, the mechanical tradition continued to flourish in physics. The theories of electric and magnetic fields developed in the nineteenth century by Michael Faraday and James Clerk Maxwell were couched in a mechanical framework, in terms of tensions within a pervasive physical medium, often called the ether. Nineteenth-century physicists were not behaving foolishly—all physicists need some sort of tentative worldview to make progress, and the mechanical worldview seemed as good a candidate as any. But it survived too long.

The final turn away from mechanism in electromagnetic theory should have come in 1905, when Einstein's special theory of relativity in effect banished the ether and replaced it with empty space as the medium that carries electromagnetic impulses. But even then the mechanical worldview lingered on among an older generation of physicists, like the fictional Professor Victor Jakob in Russell McCormmach's poignant novel,

Night Thoughts of a Classical Physicist, who were unable to absorb the new ideas.

Mechanism had also been propagated beyond the boundaries of science and survived there to give later trouble to scientists. In the nineteenth century the heroic tradition of mechanism was incorporated, unhappily, into the dialectical materialism of Marx and Engels and their followers. Lenin, in exile in 1908, wrote a turgid book about materialism, and, although for him it was mostly a device by which to attack other revolutionaries, odds and ends of his commentary were made holy writ by his followers, and for a while dialectical materialism stood in the way of the acceptance of general relativity in the Soviet Union. As late as 1961 the distinguished Russian physicist Vladimir Fock felt compelled to defend himself from the charge that he had strayed from philosophical orthodoxy. The preface to his treatise "The Theory of Space, Time, and Gravitation" contains the remarkable statement, "The philosophical side of our views on the theory of space, time and gravitation was formed under the influence of the philosophy of dialectical materialism, in particular, under the influence of Lenin's materialism and empirical criticism."

Nothing in the history of science is ever simple. Although after Einstein there was no place in serious physics research for the old naive mechanical worldview, some elements of this view were retained in the physics of the first half of the twentieth century. On one hand, there were material particles, like the electrons, protons, and neutrons that make up ordinary matter. On the other, there were fields, such as the electric and magnetic and gravitational fields, which are produced by particles and exert forces on particles. Then in 1929 physics began to turn toward a more unified worldview. Werner Heisenberg and Wolfgang Pauli described both particles and forces as manifestations of a deeper level of reality, the level of the quantum fields. Quantum mechanics had several years earlier been applied to the electric and magnetic fields and had been used to justify Einstein's idea of particles of light, the photons. Now Heisenberg and Pauli were supposing that not only photons but all particles are bundles of the energy in various fields. In this *quantum field theory* electrons are bundles of the energy of the electron field; neutrinos are bundles of the energy of the neutrino field; and so on.

Despite this stunning synthesis, much of the work on photons and electrons in the 1930s and 1940s was set in the context of the old dualistic quantum electrodynamics, in which photons were seen as bundles of energy of the electromagnetic field but electrons were merely particles of matter. As far as electrons and photons are concerned this gives the same results as quantum field theory. But by the time that I was a graduate student in the 1950s quantum field theory had become almost universally accepted as the proper framework for fundamental physics. In the physicist's recipe for the world the list of ingredients no longer included particles, but only a few kinds of fields.

From this story we may draw the moral that it is foolhardy to assume that one knows even the terms in which a future final theory will be formulated. Richard Feynman once complained that journalists ask about future theories in terms of the ultimate particle of matter or the final unification of all the forces, although in fact we have no idea whether these are the right questions. It seems unlikely that the old naive mechanical worldview will be resurrected or that we will have to return to a dualism of particles and fields, but even quantum field theory is not secure. There are difficulties in bringing gravitation into the framework of quantum field theory. In the effort to overcome these difficulties there has recently emerged a candidate for a final theory in which quantum fields are themselves just low-energy manifestations of glitches in space-time known as strings. We are not likely to know the right questions until we are close to knowing the answers.

Although naive mechanism seems safely dead, physics continues to be troubled by other metaphysical presuppositions, particularly those having to do with space and time. Duration in time is the only thing we can measure (however imperfectly) by thought alone, with no input from our senses, so it is natural to imagine that we can learn something about the dimension of time by pure reason. Kant taught that space and time are not part of external reality but are rather preexisting structures in our minds that allow us to relate objects and events. To a Kantian the most shocking thing about Einstein's theories was that they demoted space and time to the status of ordinary aspects of the physical universe, aspects that could be affected by motion (in special relativity) or gravitation (in general relativity). Even now, almost a century after the advent of special relativity, some physicists still think that there are things that can be said about space and time on the basis of pure thought.

This intransigent metaphysics comes to the surface especially in discussions of the origin of the universe. According to the standard big-bang theory the universe came into existence in a moment of infinite temperature and density some ten to fifteen billion years ago. Again and again when I have given a talk about the big-bang theory someone in the audience during the question period has argued that the idea of a beginning is absurd; whatever moment we say saw the beginning of the big bang, there must have been a moment before that one. I have tried to explain that this is not necessarily so. It is true for instance that in our ordinary experience however cold it gets it is always possible for it to get colder, but there is such a thing as absolute zero; we cannot reach temperatures below absolute zero not because we are not sufficiently clever but because temperatures below absolute zero simply have no meaning. Stephen

Hawking has offered what may be a better analogy; it makes sense to ask what is north of Austin or Cambridge or any other city, but it makes no sense to ask what is north of the North Pole. Saint Augustine famously wrestled with this problem in his *Confessions* and came to the conclusion that it is wrong to ask what there was before God created the universe, because God, who is outside time, created time along with the universe. The same view was held by Moses Maimonides.

I should acknowledge here that in fact we do not know if the universe did begin at a definite time in the past. Andre Linde and other cosmologists have recently presented plausible theories that describe our present expanding universe as just a small bubble in an infinitely old megauniverse, in which such bubbles are eternally appearing and breeding new bubbles. I am not trying here to argue that the universe undoubtedly has some finite age, only that it is not possible to say on the basis of pure thought that it does not.

Here again, we do not even know that we are asking the right questions. In the latest version of string theories space and time arise as derived quantities, which do not appear in the fundamental equations of the theory. In these theories space and time have only an approximate significance; it makes no sense to talk about any time closer to the big bang than about a million trillion trillion trillionth of a second. In our ordinary lives we can barely notice a time interval of a hundredth of a second, so the intuitive certainties about the nature of time and space that we derive from our everyday experience are not really of much value in trying to frame a theory of the origin of the universe.

It is not in metaphysics that modern physics meets its greatest troubles, but in epistemology, the study of the nature and sources of knowledge. The epistemological doctrine of positivism (or in some versions logical positivism) demands not only that science must ultimately test its theories against observation (which is hardly in doubt) but that every aspect of our theories must at every point refer to observable quantities. That is, although physical theories may involve aspects that have not yet been studied observationally and would be too expensive to study this year or next year, it would be inadmissible for our theories to deal with elements that could not in principle ever be observed. A great deal is at stake here, because positivism if valid would allow us to discover valuable clues about the ingredients of the final theory by using thought experiments to find out what sorts of things can in principle be observed.

The figure most often associated with the introduction of positivism into physics is Ernst Mach, physicist and philosopher of fin-de-siècle Vienna, for whom positivism served largely as an antidote to the metaphysics of Immanuel Kant. Einstein's 1905 paper on special relativity shows the obvious influence of Mach; it is full of observers measuring distances and times with rulers, clocks, and rays of light. Positivism helped to free Einstein from the notion that there is an absolute sense to a statement that two events are simultaneous; he found that no measurement could provide a criterion for simultaneity that would give the same result for all observers. This concern with what can actually be observed is the essence of positivism. Einstein acknowledged his debt to Mach; in a letter to him a few years later, he called himself "your devoted student." After the First World War, positivism was further developed by Rudolf Carnap and the members of the Vienna Circle of philosophers, who aimed at a reconstruction of science along philosophically satisfactory lines, and did succeed in clearing away much metaphysical rubbish.

Positivism also played an important part in the birth of modern quantum mechanics. Heisenberg's great first paper on quantum mechanics in 1925 starts with the observation that "it is well known that the formal rules which are used in [the 1913 quantum theory of Bohr] for calculating observable quantities such as the energy of the hydrogen atom may be seriously criticized on the grounds that they contain, as basic elements, relationships between quantities that are apparently unobservable in principle, e.g., position and speed of revolution of the electron." In the spirit of positivism, Heisenberg admitted into his version of quantum mechanics only observables, such as the rates at which an atom might spontaneously make a transition from one state to another by emitting a quantum of radiation. The uncertainty principle, which is one of the foundations of the probabilistic interpretation of quantum mechanics, is based on Heisenberg's positivistic analysis of the limitations we encounter when we set out to observe a particle's position and momentum.

Despite its value to Einstein and Heisenberg, positivism has done as much harm as good. But, unlike the mechanical worldview, positivism has preserved its heroic aura, so that it survives to do damage in the future. George Gale even blames positivism for much of the current estrangement between physicists and philosophers.

Positivism was at the heart of the opposition to the atomic theory at the turn of the twentieth century. The nineteenth century had seen a wonderful refinement of the old idea of Democritus and Leucippus that all matter is composed of atoms, and the atomic theory had been used by John Dalton and Amadeo Avogadro and their successors to make sense out of the rules of chemistry, the properties of gases, and the nature of heat. Atomic theory had become part of the ordinary language of physics and chemistry. Yet the positivist followers of Mach regarded this as a departure from the proper procedure of science because these atoms could not be observed with any technique that was then imaginable. The positivists decreed that scientists should concern themselves with reporting the results of observation, as for instance that it takes 2 volumes of hydrogen to combine with 1 volume of oxygen to make water vapor, but they should not concern themselves with speculations about metaphysical ideas that this is because the water molecule consists of two atoms of hydrogen and one atom of oxygen, because they could not observe these atoms or molecules. Mach himself never made his peace with the existence of atoms. As late as 1910, after atomism had been accepted by nearly everyone else, Mach wrote in a running debate with Planck that, "if belief in the reality of atoms is so crucial, then I renounce the physical way of thinking. I will not be a professional physicist, and I hand back my scientific reputation."

The resistance to atomism had a particularly unfortunate effect in retarding the acceptance of statistical mechanics, the reductionist theory that interprets heat in terms of the statistical distribution of the energies of the parts of any system. The development of this theory in the work of Maxwell, Boltzmann, Gibbs, and others was one of the triumphs of nineteenth-century science, and in rejecting it the positivists were making the worst sort of mistake a scientist can make: not recognizing success when it happens.

Positivism did harm in other ways that are less well known. There is a famous experiment performed in 1897 by J. J. Thomson, which is generally regarded as the discovery of the electron. (Thomson was Maxwell's and Rayleigh's successor as Cavendish Professor at the University of Cambridge.) For some years physicists had puzzled over the mysterious phenomenon of cathode rays, rays that are emitted when a metal plate in a glass vacuum tube is connected to the negative terminal of a powerful electric battery, and that show their presence through a luminous spot where they strike the far end of the glass tube. The picture tubes in modern television sets are nothing but cathode ray tubes in which the intensity of the rays is controlled by the signals sent out by television stations. When cathode rays were first discovered in the nineteenth century no one at first knew what they were. Then Thomson measured the way the cathode rays are bent by electric and magnetic fields as they pass through the vacuum tube. It turned out that the amount of bending of these rays was consistent with the hypothesis that they are made up of particles that carry a definite quantity of electric charge and a definite quantity of mass, always in the same ratio of mass to charge. Because the mass of these particles turned out to be so much smaller than the masses of atoms, Thomson leapt to the conclusion that these particles are the fundamental constituents of atoms and the carriers of electric charge in all currents of electricity, in wires and atoms as well as in cathode-ray tubes. For this, Thomson regarded himself, and has become universally regarded by historians, as the discoverer of a new form of matter, a particle for which he picked up a name that was already current in the theory of electrolysis: the electron.

Yet the same experiment was done in Berlin at just about the same time by Walter Kaufmann. The main difference between Kaufmann's experiment and Thomson's was that Kaufmann's was better. It yielded a result for the ratio of the electron's charge and mass that today we know was more accurate than Thomson's. Yet Kaufmann is never listed as a discoverer of the electron, because he did not think that he had discovered a new particle. Thomson was working in an English tradition going back to Newton, Dalton, and Prout—a tradition of speculation about atoms and their constituents. But Kaufmann was a positivist; he did not believe that it was the business of physicists to speculate about things that they could not observe. So Kaufmann did not report that he had discovered a new kind of particle, but only that whatever it is that is flow-ing in a cathode ray, it carries a certain ratio of electric charge to mass.

The moral of this story is not merely that positivism was bad for Kaufmann's career. Thomson, guided by his belief that he had discovered a fundamental particle, went on and did other experiments to explore its properties. He found evidence of particles with the same ratio of mass to charge emitted in radioactivity and from heated metals, and he carried out an early measurement of the electric charge of the electron. This measurement, together with his earlier measurement of the ratio of charge to mass, provided a value for the mass of the electron. It is the sum of all these experiments that really validates Thomson's claim to be the discoverer of the electron, but he would probably never have done them if he had not been willing to take seriously the idea of a particle that at that time could not be directly observed.

In retrospect the positivism of Kaufmann and the opponents of atomism seems not only obstructive but also naive. What after all does it mean to observe anything? In a narrow sense, Kaufmann did not even observe the deflection of cathode rays in a given magnetic field; he measured the position of a luminous spot on the downstream side of the vacuum tube when wires were wound a certain number of times around a piece of iron near the tube and connected to a certain electric battery and used accepted theory to interpret this in terms of ray trajectories and magnetic fields. Very strictly speaking, he did not even do that: he experienced certain visual and tactile sensations that he interpreted in terms of luminous spots and wires and batteries. It has become a commonplace among historians of science that observation can never be freed of theory.

The final surrender of the anti-atomists is usually taken to be a statement by the chemist Wilhelm Ostwald in the 1908 edition of his *Outlines of General Chemistry:* "I am now convinced that we have recently become possessed of experimental evidence of the discrete or grained nature of matter, which the atomic hypothesis sought in vain for hundreds and thousands of years." The experimental evidence that Ostwald quoted consisted of measurements of molecular impacts in the so-called Brownian motion of tiny particles suspended in liquids, together with Thomson's measurement of the charge of the electron. But if one understands how theory-laden are all experimental data, it becomes apparent that all the successes of the atomic theory in chemistry and statistical mechanics already in the nineteenth century had constituted an observation of atoms.

Heisenberg himself records that Einstein had second thoughts about the positivism of his initial approach to relativity. In a lecture in 1974 Heisenberg recalled a conversation he had with Einstein in Berlin in early 1926:

> I pointed out [to Einstein] that we cannot, in fact, observe such a path [of an electron in an atom]; what we actually record are frequencies of the light radiated by the atom, intensities and transition probabilities, but no actual path. And since it is but rational to introduce into a theory only such quantities as can be directly observed, the concept of electron paths ought not, in fact, to figure in the theory. To my astonishment, Einstein was not at all satisfied with this argument. He thought that every theory in fact contains unobservable quantities. The principle of employing only observable quantities simply cannot be consistently carried out. And when I objected that in this I had merely been applying the type of philosophy that he, too, had made the basis of his special theory of relativity, he answered simply: "Perhaps I did use such philosophy earlier, and also wrote it, but it is nonsense all the same."

Even earlier, in a Paris lecture in 1922, Einstein referred to Mach as "un bon mécanicien" but a "deplorable philosophe."

Despite the victory of atomism and the defection of Einstein, the theme of positivism has continued to be heard from time to time in the physics of the twentieth century. The positivist concentration on observables like particle positions and momenta has stood in the way of a "realist" interpretation of quantum mechanics, in which the wave function is the representation of physical reality. Positivism also played a part in obscuring the problem of infinities. As we have seen, Oppenheimer in 1930 noticed that the theory of photons and electrons known as quantum electrodynamics led to an absurd result, that the emission and absorption of photons by an electron in an atom would give the atom an infinite energy. The problem of infinities worried theorists throughout the 1930s and 1940s and led to a general supposition that quantum electrodynamics simply becomes inapplicable for electrons and photons of very high energy. Much of this angst over quantum electrodynamics was tinged with a positivist sense of guilt: some theorists feared that in speaking of the values of the electric and magnetic fields at a point in space occupied by an electron they were committing the sin of introducing elements into physics that in principle cannot be observed. This was true, but worrying about it only retarded the discovery of the real solution to the problem of infinities, that the infinities cancel when one is careful about the definition of the mass and charge of the electron.

Positivism also played a key role in a reaction against quantum field theory led by Geoffrey Chew in the 1960s at Berkeley. For Chew the central object of concern in physics was the S-matrix, the table that gives the probabilities for all possible outcomes of all possible particle collisions. The S-matrix summarizes everything that is actually observable about reactions involving any number of particles. S-matrix theory goes back to work of Heisenberg and John Wheeler in the 1930s and 1940s (the "S" stands for *streung*, which is German for "scattering"), but Chew and his coworkers were using new ideas about how to calculate the S-matrix without introducing any unobservable elements like quantum fields. In the end this program failed, partly because it was simply too hard to calculate the S-matrix in this way, but above all because the path to progress in understanding the weak and strong nuclear forces turned out to lie in the quantum field theories that Chew was trying to abandon.

The most dramatic abandonment of the principles of positivism has been in the development of our present theory of quarks. In the early 1960s Murray Gell-Mann and George Zweig independently tried to reduce the tremendous complexity of the zoo of particles then known at that time. They proposed that almost all these particles are composed of a few simple (and even more elementary) particles that Gell-Mann named quarks. This idea at first did not seem at all outside the mainstream of the way that physicists were accustomed to think; it was after all one more step in a tradition that had started with Leucippus and Democritus, of trying to explain complicated structures in terms of simpler, smaller constituents. The quark picture was applied in the 1960s to a great variety of physical problems having to do with the properties of the neutrons and protons and mesons and all the other particles that were supposed to be made up out of quarks, and generally it worked quite well. Yet the best efforts of experimental physicists in the 1960s and the early 1970s proved inadequate to dislodge quarks from the particles that were supposed to contain them. This seemed crazy. Ever since Thomson pulled electrons out of atoms in a cathode ray tube, it had always been possible to break up any composite system like a molecule or an atom or a nucleus into the individual particles of which it is composed. Why then should it be impossible to isolate free quarks?

The quark picture began to make sense with the advent in the early 1970s of quantum chromodynamics, our modern theory of strong nuclear forces, which forbids any process in which a free quark might be isolated. The breakthrough came in 1973, when independent calculations by David Gross and Frank Wilczek at Princeton and David Politzer at Harvard showed that certain kinds of quantum field theory have a peculiar property known as "asymptotic freedom," that the forces in these theories decrease at high energies. Just such a decrease in force had been observed in experiments on high-energy scattering going back to 1967, but this was the first time that any theory could be shown to have forces that behave in this way. This success rapidly led to one of these quantum field theories, the theory of quarks and gluons known as quantum chromodynamics, rapidly being accepted as the correct theory of the strong nuclear forces.

Originally it was assumed that gluons had not been observed to be produced in elementary particle collisions because they are heavy, and there had not been enough energy available in these collisions to produce the large gluon masses. Soon after the discovery of asymptotic freedom a few theorists proposed instead that the gluons are massless, like photons. If this were true, then the reason that gluons and presumably also quarks are not observed would have to be that exchange of the massless gluons between quarks or gluons produces long-range forces that make it impossible in principle to pull either quarks or gluons apart from each other. It is now believed that if you try, for instance, to pull apart a meson (a particle composed of a quark and an antiquark) the force needed increases as the quark and antiquark are pulled farther apart, until eventually you have to put so much energy into the effort that there is enough energy available to create a new quark-antiquark pair. An antiquark then pops out of the vacuum and joins itself to the original quark, while a quark pops out of the vacuum and joins itself to the original antiquark, so that instead of having a free quark and antiquark you simply have two quark-antiquark pairs—that is, two mesons. The metaphor has often been used that this is like trying to pull apart two ends of a piece of string: you can pull and pull, and eventually, if you put enough energy into the effort, the string breaks, but you do not find yourself with two isolated ends of the original piece of string; what you have are two pieces of string, each of which has two ends. The idea that quarks and gluons can in principle never be observed in isolation has become part of the accepted wisdom of modern elementary particle physics, but it does not stop us from describing neutrons and protons and mesons as composed of quarks. I cannot imagine anything that Ernst Mach would like less.

The quark theory was only one step in a continuing process of reformulation of physical theory in terms that are more and more fundamental and at the same time farther and farther from everyday experience. How can we hope to make a theory based on observables when no aspect of our experience—perhaps not even space and time—appears at the most fundamental level of our theories? It seems to me unlikely that the positivist attitude will be of much help in the future.

Metaphysics and epistemology have at least been intended to play a constructive role in science. In recent years science has come under attack from unfriendly commentators joined under the banner of relativism. The philosophical relativists deny the claim of science to the discovery of objective truth; they see it as merely another social phenomenon, not fundamentally different from a fertility cult or a potlatch.

Philosophical relativism stems in part from the discovery by philosophers and historians of science that there is a large subjective element in the process by which scientific ideas become accepted. We have seen here the role that aesthetic judgments play in the acceptance of new physical theories. This much is an old story to scientists (though philosophers and historians sometimes write as if we were utterly naive about this). In his celebrated book *The Structure of Scientific Revolutions* Thomas Kuhn went a step further and argued that in scientific revolutions the standards (or "paradigms") by which scientists judge theories change, so that the new theories simply cannot be judged by the prerevolutionary standards. There is much in Kuhn's book that fits my own experience in science. But in the last chapter Kuhn tentatively attacked the view that science makes progress toward objective truths: "We may, to be more precise, have to relinquish the notion, explicit or implicit, that changes of paradigm carry scientists and those who learn from them closer and closer to the truth." Kuhn's book lately seems to have become read (or at least quoted) as a manifesto for a general attack on the presumed objectivity of science.

There has as well been a growing tendency starting with the work of Robert Merton in the 1930s for sociologists and anthropologists to treat the enterprise of science (or at least, sciences other than sociology and anthropology) by the same methods that are used to study other social phenomena. Science *is* of course a social phenomenon, with its own reward system, its revealing snobberies, its interesting patterns of alliance and authority. For instance, Sharon Traweek has spent years with elementary particle experimentalists at both the Stanford Linear Accelerator Center and the KEK Laboratory in Japan and has described what she had seen from the perspective of an anthropologist. This kind of big science is a natural topic for anthropologists and sociologists, because scientists belong to an anarchic tradition that prizes individual initiative, and yet they find in today's experiments that they have to work together in teams of hundreds. As a theorist I have not worked in such a team, but many of her observations seem to me to have the ring of truth, as for instance:

> The physicists see themselves as an elite whose membership is determined solely by scientific merit. The assumption is that everyone has a fair start. This is underscored by the rigorously informal dress code, the similarity of their offices, and the "first

naming" practices in the community. Competitive individualism is considered both just and effective: the hierarchy is seen as a meritocracy which produces fine physics. American physicists, however, emphasize that science is not democratic: decisions about scientific purposes should not be made by majority rule within the community, nor should there be equal access to a lab's resources. On both these issues, most Japanese physicists assume the opposite.

In the course of such studies, sociologists and anthropologists have discovered that even the process of change in scientific theory is a social one. A recent book on peer review remarks that "scientific truths are, at bottom, widely quoted social agreements about what is 'real', arrived at through a distinctively 'scientific process' of negotiation." Close observation of scientists at work at the Salk Institute led the French philosopher Bruno Latour and the English sociologist Steve Woolgar to comment, "The negotiations as to what counts as a proof or what constitutes a good assay are no more or less disorderly than any argument between lawyers and politicians."

It seems to have been an easy step from these useful historical and sociological observations to the radical position that the content of the scientific theories that become accepted is what it is because of the social and historical setting in which the theories are negotiated. (The elaboration of this position is sometimes known as the strong program in the sociology of science.) This attack on the objectivity of scientific knowledge is made explicit and even brought into the title of a book by Andrew Pickering: *Constructing Quarks*. In his final chapter, he comes to the conclusion: "And, given their extensive training in sophisticated mathematical techniques, the preponderance of mathematics in particle physicists' accounts of reality is no more hard to explain than the fondness of ethnic groups for their native language. On the view advocated in this chapter, there is no obligation upon anyone framing a view of the world to take account of what twentieth-century science has to say." Pickering describes in detail a great change of emphasis in high-energy experimental physics that took place in the late 1960s and early 1970s. Instead of a commonsense (Pickering's term) approach of concentrating on the most conspicuous phenomena in collisions of high-energy particles (i.e., the fragmentation of the particles into great numbers of other particles going mostly in the original direction of the particle beams), experimentalists instead began to do experiments suggested by theorists, experiments that focused on rare events, such as those in which some high-energy particle emerges from the collision at a large angle to the incoming beam direction.

There certainly was a change of emphasis in high-energy physics, pretty much as described by Pickering, but it was driven by the necessities of the historical mission of physics. A proton consists of three quarks, together with a cloud of continually appearing and disappearing gluons and quark-antiquark pairs. In most collisions between protons the energy of the initial particles goes into a general disruption of these clouds of particles, like a collision of two garbage trucks. These may be the most conspicuous collisions, but they are too complicated to allow us to calculate what should happen according to our current theory of quarks and gluons and so they are useless for testing that theory. Every once in a while, however, a quark or gluon in one of the two protons hits a quark or gluon in the other proton head on, and their energy becomes available to eject these quarks or gluons at high energy from the debris of the collision, a process whose rate we do know how to calculate. Or the collision may create new particles, like the W and Z particles that carry the weak nuclear force, which need to be studied to learn more about the unification of the weak and electromagnetic forces. It is these rare events that today's experiments are designed to detect. Yet Pickering, who as far as I can tell understands this theoretical background very well, still describes this change of emphasis in high-energy physics in terms suggestive of a mere change of fashion, like the shift from impressionism to cubism, or from short skirts to long.

It is simply a logical fallacy to go from the observation that science is a social process to the conclusion that the final product, our scientific theories, is what it is because of the social and historical forces acting in this process. A party of mountain climbers may argue over the best path to the peak, and these arguments may be conditioned by the history and social structure of the expedition, but in the end either they find a good path to the peak or they do not, and when they get there they know it. (No one would give a book about mountain climbing the title *Constructing Everest*.) I cannot prove that science is like this, but everything in my experience as a scientist convinces me that it is. The "negotiations" over changes in scientific theory go on and on, with scientists changing their minds again and again in response to calculations and experiments, until finally one view or another bears an unmistakable mark of objective success. It certainly feels to me that we are discovering something real in physics, something that is what it is without any regard to the social or historical conditions that allowed us to discover it.

Where then does this radical attack on the objectivity of scientific knowledge come from? One source I think is the old bugbear of positivism, this time applied to the study of science itself. If one refuses to talk about anything that is not directly observed, then quantum field theories or principles of symmetry or more generally laws of nature cannot be taken seriously. What philosophers and sociologists and anthropologists *can* study is the actual behavior of real scientists, and this behavior never follows any simple description in terms of rules of inference. But scientists have the direct experience of scientific theories as desired yet elusive goals, and they become convinced of the reality of these theories.

There may be another motivation for the attack on the realism and objectivity of science, one that is less high-minded. Imagine if you will an anthropologist who studies the cargo cult on a Pacific island. The islanders believe that they can bring back the cargo aircraft that made them prosperous during World War II by building wooden structures that imitate radar and radio antennas. It is only human nature that this anthropologist and other sociologists and anthropologists in similar circumstances would feel a frisson of superiority, because they know as their subjects do not that there is no objective reality to these beliefs—no cargo-laden C-47 will ever be attracted by the wooden radars. Would it be surprising if, when anthropologists and sociologists turned their attention to studying the work of scientists, they tried to recapture that delicious sense of superiority by denying the objective reality of the scientists' discoveries?

Relativism is only one aspect of a wider, radical, attack on science itself. Feyerabend called for a formal separation of science and society like the separation of church and state, reasoning that "science is just one of the many ideologies that propel society and it should be treated as such." The philosopher Sandra Harding calls modern science (and especially physics) "not only sexist but also racist, classist, and culturally coercive" and argues, "Physics and chemistry, mathematics and logic, bear the fingerprints of their distinctive cultural creators no less than do anthropology and history." Theodore Roszak urges that we change "the fundamental sensibility of scientific thought . . . even if we must drastically revise the professional character of science and its place in our culture."

These radical critics of science seem to be having little or no effect on the scientists themselves. I do not know of any working scientist who takes them seriously. The danger they present to science comes from their possible influence on those who have not shared in the work of science but on whom we depend, especially on those in charge of funding science and on new generations of potential scientists. Recently the minister in charge of government spending on civil science in Britain was quoted by *Nature* as speaking with approval of a book by Bryan Appleyard that has as its theme that science is inimical to the human spirit.

I suspect that Gerald Holton is close to the truth in seeing the radical attack on science as one symptom of a broader hostility to Western civilization that has bedeviled Western intellectuals from Oswald Spengler on. Modern science is an obvious target for this hostility; great art and literature have sprung from many of the world's civilizations, but ever since Galileo scientific research has been overwhelmingly dominated by the West.

This hostility seems to me to be tragically misdirected. Even the most frightening Western applications of science such as nuclear weapons represent just one more example of mankind's timeless efforts to destroy itself with whatever weapons it can devise. Balancing this against the benign applications of science and its role in liberating the human spirit, I think that modern science, along with democracy and contrapuntal music, is something that the West has given the world in which we should take special pride.

In the end this issue will disappear. Modern scientific methods and knowledge have rapidly diffused to non-Western countries like Japan and India and indeed are spreading throughout the world. We can look forward to the day when science can no longer be identified with the West but is seen as the shared possession of humankind.

CHAPTER VIII

TWENTIETH CENTURY BLUES

Blues,
Twentieth Century Blues,
Are getting me down.
Who's
Escaped those weary
Twentieth Century Blues.

Noël Coward, *Cavalcade*

Wherever we have been able to pursue our chains of questions about force and matter sufficiently far, the answers have been found in the standard model of elementary particles. And at every high-energy physics conference since the late 1970s, experimentalists have reported increasingly precise agreement between their results and the predictions of the standard model. You might think that high-energy physicists would be feeling some sense of contentment, so why are we so blue?

First of all, the standard model describes the electromagnetic and weak and strong nuclear forces, but it leaves out a fourth force, actually the first known of all the forces, the force of gravitation. This omission is not simply absentmindedness; as we shall see, there are formidable mathematical obstacles to describing gravitation in the same language that we use to describe the other forces in the standard model, the language of quantum field theory. Second, although the strong nuclear force *is* included in the standard model, it appears as something quite different from the electromagnetic and weak nuclear forces, not as part of a unified picture. Third, although the standard model treats the electromagnetic and the weak nuclear forces in a unified way, there are obvious differences between these two forces. (For instance, under ordinary circumstances the weak nuclear force is much weaker than the electromagnetic force.) We have a general idea of how the differences between the electromagnetic and weak forces arise, but we do not fully understand the source of these differences. Finally, apart from the problem of unifying the four forces, the standard model involves many features that are not dictated by fundamental principles (as we would like) but instead simply have to be taken from experiment. These apparently arbitrary features include a menu of particles, a number of constants such as ratios of masses, and even the symmetries themselves. We can easily imagine that any or all of these features of the standard model might have been different.

To be sure, the standard model is a huge improvement over the mishmash of approximate symmetries, ill-formulated dynamical assumptions, and mere facts that my generation of physicists had to learn in graduate school. But the standard model is clearly not the final answer, and to go beyond it we shall have to grapple with all its failings.

All these problems with the standard model touch in one way or another on a phenomenon known as *spontaneous symmetry breaking*. The discovery of this phenomenon has been one of the great liberating developments of twentieth-century science, first in condensed matter physics and then in the physics of elementary particles. Its largest success has been in explaining the differences between the weak and electromagnetic forces, so the electroweak theory will be a good place for us to start to take a look at the phenomenon of spontaneous symmetry breaking.

The electroweak theory is the part of the standard model that deals with weak and electromagnetic forces. It is based on an *exact* symmetry principle, which says that the laws of nature take the same form if everywhere in the equations of the theory we replace the fields of the electron and neutrino with mixed fields—for instance, one field that is 30% electron and 70% neutrino and another field that is 70% electron and 30% neutrino—and at the same time similarly mix up the fields of other families of particles, such as the up quark and the down quark. This symmetry principle is called *local,* meaning that the laws of nature are supposed to be unchanged even if these mixtures vary from moment to moment or from one location to another. There is one other family of fields, whose existence is *dictated* by this symmetry principle in something like the way that the existence of the gravitational field is dictated by the symmetry among different coordinate systems. This family consists of the fields of the photon and the W and Z particles, and these fields too must be mixed up with each other when we mix up the electron and neutrino fields and the quark fields. The exchange of photons is responsible for the electromagnetic force, while the exchange of W and Z particles produces the weak nuclear force, so this symmetry between electrons and neutrinos is also a symmetry between the electromagnetic force and the weak nuclear force.

But this symmetry is certainly not manifest in nature, which is why it took so long to be discovered. For example, electrons and W and Z particles have masses, but neutrinos and photons do not. (It is the large mass of the W and Z particles that makes the weak forces so much weaker than the electromagnetic forces.) In other words, the symmetry that relates the electron and neutrino and so on is a property of the underlying equations of the standard model, equations that dictate the properties of the elementary particles, but this symmetry is not satisfied by the *solutions* of these equations—the properties of the particles themselves.

To see how equations can have a symmetry when their solutions do not, suppose that our equations were completely symmetric between two types of particle, such as the up quark and the down quark, and that we wished to solve these equations to find the masses of the two particles. One might suppose that the symmetry between the two quark types would dictate that the two masses should turn out to be equal, but this is not the only possibility. The symmetry of the equations does not rule out the possibility that the solution might turn out to give the up quark a mass greater than the down quark mass; it requires only that, in that case, there will be a *second* solution of the equations in which the down quark mass is greater than the up quark mass and by exactly the same amount. That is, the symmetry of the equations is not necessarily reflected in each individual solution of these equations, but only in the pattern of *all* the solutions of these equations. In this simple example the actual properties of the quarks would correspond to one or the other of the two solutions, representing a breakdown in the symmetry of the underlying theory. Notice that it does not really matter which of the two solutions is realized in nature— if the only difference between the up and down quarks were in their masses, then the difference between the two solutions would be simply a matter of which quarks we choose to call up and down. Nature as we know it represents one solution of all the equations of the standard model, and it makes no difference *which* solution, as long as these different solutions are all related by exact symmetry principles.

In such cases we say that the symmetry is broken, although a better term would be "hidden," because the symmetry is still there in the equations, and these equations govern the properties of the particles. We call this phenomenon a *spontaneous symmetry breaking* because nothing breaks the symmetry in the equations of the theory; the symmetry breaking appears spontaneously in the various solutions of these equations.

It is principles of symmetry that give our theories much of their beauty. That is why it was so exciting when elementary particle physicists started to think about spontaneous symmetry breaking in the early 1960s. It suddenly came home to us that there is much more symmetry in the laws of nature than one would guess merely by looking at the properties of elementary particles. Broken symmetry is a very Platonic notion: the reality we observe in our laboratories is only an imperfect reflection of a deeper and more beautiful reality, the reality of the equations that display all the symmetries of the theory.

An ordinary permanent magnet provides a good realistic example of a broken symmetry. (This example is particularly appropriate because spontaneous symmetry breaking made its first appearance in quantum physics in Heisenberg's 1928 theory of permanent magnetism.) The equations that govern the iron atoms and magnetic field in a magnet are perfectly symmetrical with regard to directions in space; nothing in these equations distinguishes north from south or east or up. Yet, when a piece of iron is cooled below 770°C, it spontaneously develops a magnetic field pointing in some specific direction, breaking the symmetry among different directions. A race of little beings who are born and live their whole lives inside a permanent magnet would take a long time to realize that the laws of nature actually possess a symmetry regarding the different directions in space, and that there seems to be a preferred direction in their environment only because the spins of the iron atoms have spontaneously lined up in the same direction, producing a magnetic field.

We, like the beings in the magnet, have recently discovered

a symmetry that happens to be broken in *our* universe. It is the symmetry relating the weak and electromagnetic forces, whose breakdown is shown, for instance, by the dissimilarities between the massless photon and the very heavy W and Z particles. One great difference between the symmetry breakdown in the standard model and in a magnet is that the origin of magnetization is well understood. It occurs because of known electromagnetic forces between neighboring iron atoms that tend to make their spins line up parallel to each other. The standard model is more of a mystery. None of the known forces of the standard model is strong enough to be responsible for the observed breaking of the symmetry between the weak and electromagnetic forces. The most important thing that we still do not know about the standard model is just what causes the breaking of the electroweak symmetry.

In the original version of the standard theory of weak and electromagnetic forces, the breakdown of the symmetry between these forces was attributed to a new field that was introduced into the theory for just this purpose. This field was supposed to turn on spontaneously like the magnetic field in a permanent magnet, pointing in some definite direction—not a direction in ordinary space, but rather a direction on the imaginary little dials that distinguish electrons from neutrinos, photons from W and Z particles, and so on. The value of the field that breaks the symmetry is commonly called its *vacuum value*, because the field takes this value in the vacuum, far from the influence of any particles. After a quarter century we still do not know whether this simple picture of symmetry breaking is correct, but it remains the most plausible possibility.

This is not the first time that physicists have proposed the existence of a new field or particle in order to satisfy some theoretical requirement. Physicists in the early 1930s were worried about an apparent violation of the law of conservation of energy when a radioactive nucleus undergoes the process known as beta decay. In 1932 Wolfgang Pauli proposed the existence of a convenient particle he called the neutrino, in order to account for the energy that was observed to be lost in this process. The elusive neutrino was eventually discovered experimentally, over two decades later. Proposing the existence of something that has not yet been observed is a risky business, but it sometimes works.

Like any other field in a quantum-mechanical theory, this new field that is responsible for electroweak symmetry breaking would have an energy and momentum that come in bundles, known as quanta. The electroweak theory tells us that at least one of these quanta should be observable as a new elementary particle. Several years before Salam and I developed a theory of weak and electromagnetic forces based on spontaneous symmetry breaking, the mathematics of simpler examples of this sort of symmetry breaking had been described by a number of theorists, most clearly in 1964 by Peter Higgs of the University of Edinburgh. So the new particle that is necessary in the original version of the electroweak theory has come to be known as a *Higgs particle*.

No one has discovered a Higgs particle, but this does not yet contradict the theory; a Higgs particle could not have been seen in any experiment done so far if its mass were greater than about fifty times the proton's mass, which it well might be. (The electroweak theory is unfortunately silent about the mass of the Higgs particle, except to tell us that almost certainly it would not be heavier than a trillion volts, a thousand times the proton mass.) We need experiment to tell us whether there actually is a Higgs particle, or perhaps several Higgs particles, and to supply us with their masses.

These questions have an importance that goes beyond the question of how the electroweak symmetry is broken. One of the new things that we learned from the electroweak theory is that all the particles of the standard model, aside from the Higgs particle, get their masses from the breaking of the symmetry between the weak and electromagnetic forces. If we could somehow turn off this symmetry breaking then the electron and the W and Z particles and all the quarks would be massless, like the photon and neutrinos. The problem of understanding the masses of the known elementary particles is therefore a part of the problem of understanding the mechanism by which the electroweak symmetry is spontaneously broken. In the original version of the standard model, the Higgs particle is the only particle whose mass appears directly in the equations of the theory; the breaking of the electroweak symmetry gives all the other particles masses that are proportional to the Higgs particle mass. But we have no evidence that matters are so simple.

The problem of the cause of electroweak symmetry breaking is important not only in physics but also in our efforts to understand the early history of our universe. Just as the magnetization of a piece of iron can be wiped out and the symmetry between different directions restored by raising the temperature of the iron above 770 degrees, so also the symmetry between the weak and electromagnetic forces could be restored if we could raise the temperature of our laboratory above a few million billion degrees. At such temperatures the symmetry would be no longer hidden, but clearly apparent in the properties of the particles of the standard model. (E.g., at these temperatures the electron and the W and Z particles and all the quarks would be massless.) Temperatures like a million billion degrees cannot be created in the laboratory and do not exist today even in the centers of the hottest stars. But according to the simplest version of the generally accepted big-bang theory of cosmology, there was a moment some ten to twenty billion years in the past when the temperature of the universe was infinite. About one ten-billionth of a second after this starting moment the temperature of the universe had dropped to a few million billion degrees, and at this time the symmetry between the weak and electromagnetic forces became broken.

This symmetry breaking probably did not happen instantaneously and uniformly. In more familiar "phase transitions" like the freezing of water or the magnetization of iron the transition may occur a little earlier or later at one point or another and may not occur in the same way everywhere, as we see for instance in the formation of separate small crystals of ice or of domains in a magnet in which the magnetization points in different directions. This sort of complication in the electroweak phase transition might have had various detectable effects, for instance on the abundances of the light elements that were formed a few minutes later. But we cannot assess these possibilities until we learn the mechanism by which the electroweak symmetry became broken.

We know that there is a broken symmetry between the weak and electromagnetic forces because the theory that is based on this symmetry *works*—it makes a large number of successful predictions about the properties of the W and Z particles and about the forces they transmit. But we are not actually sure that the electroweak symmetry is broken by the vacuum value of some field in the theory or that there is a Higgs particle. *Something* has to be included in the electroweak theory to break this symmetry, but it is possible that the breakdown of the electroweak symmetry is due to indirect effects of some new kind of extra-strong force that does not act on ordinary quarks or electrons or neutrinos and for that reason has not yet been detected. Such theories were developed in the late 1970s but have problems of their own. One of the chief missions of the Superconducting Super Collider now under construction is to settle this issue.

This is not the end of the story of spontaneous symmetry breaking. The idea of spontaneous symmetry breaking has also played a part in our efforts to bring the third force of the standard model, the strong nuclear force, into the same unified framework as the weak and electromagnetic forces. Although the obvious differences between the weak and electromagnetic forces are explained in the standard model as a result of spontaneous symmetry breaking, this is not true of the strong nuclear force; there is no symmetry even in the equations of the standard model that relates the strong nuclear forces to the weak and electromagnetic forces. Beginning in the early 1970s, this led to a search for a theory underlying the standard model, in which the strong as well as the weak and electromagnetic interactions would be unified by a single large and spontaneously broken group of symmetries.

There was an obvious obstacle to any sort of unification along these lines. The apparent strengths of the forces in any field theory depend on two kinds of numerical parameter: the masses (if any) of the particles like W and Z particles that transmit the forces, and certain *intrinsic strengths* (also known as coupling constants) that characterize the likelihood for particles like photons or gluons or W and Z particles to be emitted and reabsorbed in particle reactions. The masses arise from spontaneous symmetry breaking, but the intrinsic strengths are numbers that appear in the underlying equation of the theory. Any symmetry that connects the strong with the weak and electromagnetic forces, even if spontaneously broken, would dictate that the intrinsic strengths of the electroweak and strong forces should (with suitable conventions for how they are defined) all be equal. The apparent differences between the strengths of the forces would have to be attributed to the spontaneous symmetry breaking that produces differences in the masses of the particles that transmit the forces, in much the same way that the differences between the electromagnetic and weak forces arise in the standard model from the fact that the electroweak symmetry breaking gives the W and Z particles very large masses, while the photon is left massless. But it is clear that the intrinsic strengths of the strong nuclear force and the electromagnetic force are *not* equal; the strong nuclear force, as its name suggests, is much stronger than the electromagnetic force, even though both of these forces are transmitted by massless particles, the gluons and photons.

In 1974 an idea appeared that offered a way around this obstacle. The intrinsic strengths of all these forces actually depend very weakly on the energies of the processes in which they are measured. In any sort of theory that unifies the strong with the electroweak forces, these intrinsic strengths would be expected to be equal at *some* energy, but this energy might be very different from the energies of present experiments. There are three independent intrinsic strengths of the forces in the standard model (this is one of the reasons we have been dissatisfied with it as a final theory), so it is not a trivial condition that there should be *any* one energy at which all the strengths should become equal. By imposing this condition it was possible to make one prediction that related the strengths that the forces should have at the energies of existing experiments, a prediction that has turned out to be in reasonable agreement with experiment. This is just a single quantitative success, but it is enough to encourage us that there is something to these ideas.

It was also possible in this way to estimate the energy at which the intrinsic strengths of the forces become equal. The strong nuclear force is much stronger than the other forces at the energies of existing accelerators, and, according to quantum chromodynamics, it weakens only very slowly with increasing energy, so the energy where the forces of the standard model all become equally strong is predicted to be very high: it is calculated to be about a million billion billion volts. (Recent updates of this calculation suggest an energy closer to ten million billion billion volts.) If there really were some spontaneously broken symmetry that linked the strong and electroweak forces, then there would have to exist new heavy particles in order to fill out the complete family of force-carrying particles along with the W, Z, photon, and gluons. In this case, the energy of a few million billion billion volts could be identified as the energy contained in the mass of these new heavy particles. As we shall see, in today's superstring theories there is no need to assume the existence of a separate new symmetry linking the strong and electroweak forces, but it is still the case that the intrinsic strengths of the strong and electroweak forces would all become equal at some very high energy, calculated to be about ten million billion billion volts.

This may seem like just another indigestibly large number, but when the estimate of a million billion billion volts was made in 1974, it set off bells in the minds of theoretical physicists. We all knew of another very large energy, that naturally appears in any theory that attempts to unify gravitation with the other forces of nature. Under ordinary conditions the force of gravitation is very much weaker than the weak, strong, and electromagnetic forces. No one has ever observed any effect of the gravitational forces among the particles within a single atom or molecule, and there is not much hope that anyone ever will. (The only reason that gravitation seems to us like a rather strong force in our everyday lives is that the earth contains a large number of atoms, each contributing a tiny amount to the gravitational field on the earth's surface.) But according to general relativity, gravitation is produced by and acts on energy as well as mass. This is why photons that have energy but no mass are deflected by the gravitational field of the sun. At sufficiently high energy the force of gravitation between two typical elementary particles becomes as strong as any other force between them. The energy at which this happens is about a thousand million billion billion volts. This is known as the Planck energy.*

It is striking that the Planck energy is only about a hundred times larger than the energy at which the intrinsic strengths of the strong and electroweak forces become equal, even though both of these energies are enormously larger than the energies normally encountered in elementary particle physics. The fact that these two huge energies are relatively so close suggests powerfully that the breakdown of any symmetry that unites the strong and electroweak forces is only part of a more fundamental symmetry breaking: the breakdown of whatever symmetry relates gravitation to the other forces of nature. There may be no separate unified theory of strong, weak, and electromagnetic forces, but only a truly unified theory that encompasses gravitation as well as the strong, weak, and electromagnetic forces.

Unfortunately, the reason that gravitation is left out of the standard model is that it is very difficult to describe gravitation in the language of quantum field theory. We can simply apply the rules of quantum mechanics to the field equations of general relativity, but then we run into the old problem of infinities. For instance, if we try to calculate the probabilities for what happens in a collision of two gravitons (the particles that make up a gravitational field), we get perfectly sensible contributions from the exchange of one graviton between the colliding gravitons, but, if we carry our calculations a step further and take into account the exchange of two gravitons, we begin to encounter infinite probabilities. These infinities can be canceled if

---

*In 1899 Max Planck remarked in effect that this is the natural unit of energy that can be calculated from a knowledge of the speed of light, the constant later named after him, and the constant in Newton's formula for gravitational force.

we modify Einstein's field equations by putting in a new term with an infinite constant factor that cancels the first infinity, but then when we include the exchange of *three* gravitons in our calculations we encounter new infinities, which can be canceled by adding yet more terms to the field equations, and so on, until we wind up with a theory with an unlimited number of unknown constants. A theory of this sort is actually useful in calculating quantum processes at relatively low energy, where the new terms added to the field equations are negligibly small, but it loses all predictive power when we apply it to gravitational phenomena at the Planck energy. For the present the calculation of physical processes at the Planck energy is simply beyond our reach.

Of course no one is studying processes at the Planck energy experimentally (or indeed measuring any quantum gravitational process like graviton-graviton collisions at any energy), but in order for a theory to be regarded as satisfactory it not only must agree with the results of experiments that have been done but also must make predictions that are at least plausible for experiments that in principle could be done. In this respect, general relativity was for years in the same position as was the theory of weak interactions before the development of the electroweak theory in the late 1960s: general relativity works very well wherever it can be tested experimentally, but it contains internal contradictions that show that it needs modification.

The value of the Planck energy confronts us with a formidable new problem. It is not that this energy is so large—it arises in physics at such a deep level that we can suppose that the Planck energy is simply the fundamental unit of energy that appears in the equations of the final theory. The mystery is *why are all the other energies so small?* In particular, in the original version of the standard model the masses of the electron and the W and Z particles and all the quarks are proportional to the one mass that appears in the equations of the model, the mass of the Higgs particle. From what we know of the masses of the W and Z particles, we can infer that the energy in a Higgs particle mass could not be greater than about a trillion volts. But this is at least a hundred million million times smaller than the Planck energy. This also means that there is a hierarchy of symmetries: whatever symmetry unites the gravitational and strong nuclear forces with the electroweak forces is broken roughly a hundred million million times more strongly than the symmetry that unifies the weak and electromagnetic interactions. The puzzle of explaining this enormous difference in fundamental energies is therefore known in elementary particle physics today as the *hierarchy problem*.

For over fifteen years the hierarchy problem has been the worst bone in the throat of theoretical physics. Much of the theoretical speculation of recent years has been driven by the need to solve this problem. It is not a paradox—there is no reason why some energies in the fundamental equations of physics should *not* be a hundred million million times smaller than others—but it is a mystery. That is what makes it so difficult. A paradox like a murder in a locked room may suggest its own solution, but a mere mystery forces us to look for clues beyond the problem itself.

One approach to the hierarchy problem is based on the idea of a new sort of symmetry, known as *supersymmetry*, which relates particles of different spin so that they form new "super-families." In supersymmetric theories there are several Higgs particles, but the symmetry forbids the appearance of any Higgs particle masses in the fundamental equations of the theory; what we call the Higgs particle masses in the standard model would have to arise from complicated dynamical effects that break supersymmetry. In another approach mentioned earlier we give up the idea of a field whose vacuum value breaks the electroweak symmetry and attribute this symmetry breaking instead to the effects of some new extra-strong force.

Unfortunately there is so far no sign of supersymmetry or new extra-strong forces in nature. This fact is not yet a conclusive argument against these ideas; the new particles predicted by these approaches to the hierarchy problem might well be too heavy to have been produced at existing accelerator laboratories.

We expect that Higgs particles or the other new particles required by various approaches to the hierarchy problem could be discovered at a sufficiently powerful new particle accelerator, like the Superconducting Super Collider. But there is no way that any accelerator we can now imagine will be able to concentrate onto individual particles the enormous energies at which all the forces become unified. When Democritus and Leucippus speculated about atoms at Abdera, they could not guess that these atoms were a million times smaller than the grains of sand on the Aegean beaches or that 2,300 years would pass before there was direct evidence of the existence of atoms. Our speculations have now brought us to the edge of a far wider gulf: we think that all the forces of nature become united at something like the Planck energy, a million billion times larger than the highest energy reached in today's accelerators.

The discovery of this huge gulf has changed physics in ways that go beyond the hierarchy problem. For one thing, it has cast a new light on the old problem of infinities. In the standard model as in the older quantum electrodynamics, the emission and absorption of photons and other particles of unlimitedly high energy makes infinite contributions to atomic energies and other observable quantities. To deal with these infinities, the standard model was required to have the special property of being renormalizable; that is, that all infinities in the theory should be canceled by other infinities appearing in the definition of the bare masses and other constants that enter in the equations of the theory. This condition was a powerful guide in constructing the standard model; only theories with the simplest possible field equations are renormalizable. But, because the standard model leaves out gravitation, we now think that it is only a low-energy approximation to a really fundamental unified theory and that it loses its validity at energies like the Planck energy. Why then should we take seriously what it tells us about the effects of emitting and absorbing particles of unlimitedly high energy? And if we do not take this seriously, why should we demand that the standard model be renormalizable? The problem of infinities is still with us, but it is a problem for the final theory, not for a low-energy approximation like the standard model.

As a result of this reappraisal of the problem of infinities, we now think that the field equations of the standard model are not of the very simple type that would be renormalizable but that they actually contain every conceivable term that is consistent with the symmetries of the theory. But then we must explain why the old renormalizable quantum field theories like the simplest versions of quantum electrodynamics or the standard model worked so well. The reason, we think, can be traced to the fact that all the terms in the field equations, apart from the very simple renormalizable terms, necessarily appear in these equations divided by powers of something like the Planck energy. The effect of these terms on any observed physical process would then be proportional to powers of the ratio of the energy of the process to the Planck energy, a ratio perhaps as small as one part in a million billion. This is such a tiny number that naturally no such effect has been detected. In other words, the condition of renormalizability that guided our thinking from the quantum electrodynamics of the 1940s to the standard model of the 1960s and 1970s was for practical purposes the right condition, though it was imposed for reasons that no longer seem relevant.

This change in point of view has consequences of great potential importance. The standard model in its simplest renormalizable form had certain "accidental" conservation laws, over and above the really fundamental conservation laws that follow from the symmetries of special relativity and from the internal symmetries that dictate the existence of the photon, W and Z particles, and gluons. Among these accidental conservation laws are the conservation of quark number (the total number of quarks minus the total number of antiquarks) and lepton number (the total number of electrons and neutrinos and related particles, minus the total number of their antiparticles). When we list all the possible terms in the field equations that would be consistent with the fundamental symmetries of the standard model and the condition of renormalizability, we find that there is no term in the field equations that could violate these conservation laws. It is the conservation of quark and lepton number that prevents processes like the decay of the three quarks in a proton into a positron and a photon, and thus it is this conservation law that ensures the stability of ordinary matter. But we now think that the complicated nonrenormalizable terms in the field equations that could violate the conservation of quark and lepton number are really present but just very small. These small terms in the field equations would make the proton decay (e.g., into a positron and a photon or some other neutral particle), but with a very long average lifetime, originally estimated to be about a hundred million million million million million years, or perhaps a little longer or shorter. This is about as many years as there are protons in 100 tons of water, so, if this is true, then on the average roughly about one proton a year should decay in 100 tons of water. Experiments have been looking for such a proton decay without success for years, but there will soon be a facility in Japan where 10,000 tons of water will be carefully watched for flashes of light that would signal proton decays. Perhaps this experiment will see something.

Meanwhile there have lately been intriguing hints of a possible violation of the conservation of lepton number. In the standard model this conservation law is responsible for keeping the neutrinos massless, and with this conservation law violated we would expect neutrinos to have small masses, about a hundredth to a thousandth of a volt (or in other words, about one billionth the mass of an electron). This mass is much too small to have been noticed in any laboratory experiment done so far, but it could have a subtle effect, of allowing neutrinos that start out as electron-type neutrinos (i.e., members of the same family as the electron) to turn slowly into neutrinos of other types. This may explain a long-standing puzzle, that fewer neutrinos than expected are detected to be coming from the sun. Neutrinos produced in the sun's core are mostly of electron type, and the detectors used to observe them on earth are chiefly sensitive to electron-type neutrinos, so perhaps electron-type neutrinos seem to be missing because as they pass through the sun they have turned into neutrinos of other types. Experiments to test this idea with neutrino detectors of various types are now under way in South Dakota, Japan, the Caucasus, Italy, and Canada.

If we are lucky, we may yet discover definite evidence of proton decay or neutrino masses. Or perhaps existing accelerators like the Fermilab proton-antiproton collider or the electron-positron collider at CERN may yet turn up evidence of supersymmetry. But all this is moving with glacial slowness. The summary talk at a high-energy physics conference held at any time in the last decade could (and usually did) give the same wish list of possible breakthroughs. It is all very different from the really exciting times of the past, when it seemed that every month graduate students would be racing down the corridors of university physics departments to spread the news of a new discovery. It is a tribute to the fundamental importance of elementary particle physics that very bright students continue to come into the field when so little is going on.

We could be confident of breaking out of this impasse if the Superconducting Super Collider were completed. It was planned to have enough energy and intensity to settle the question of the mechanism for electroweak symmetry breaking, either by finding one or more Higgs particles or by revealing signs of new strong forces. If the answer to the hierarchy problem is supersymmetry, then that, too, would be discovered at the Super Collider. On the other hand, if new strong forces were discovered, then the Super Collider would find a rich variety of new particles with masses of a trillion volts or so, which would have to be explored before we could guess what goes on at the much higher energies where all the forces including gravitation become unified. Either way, particle physics would be moving again. The particle physicists' campaign for the Super Collider has been spurred by a sense of desperation, that only with the data from such an accelerator can we be sure that our work will continue.

CHAPTER IX

THE SHAPE OF
A FINAL THEORY

If you can look into the seeds of time,
And say which grain will grow and which will not,
Speak then to me.

William Shakespeare, *Macbeth*

The final theory may be centuries away and may turn out to be totally different from anything we can now imagine. But suppose for a moment that it was just around the corner. What can we guess about this theory on the basis of what we already know?

The one part of today's physics that seems to me likely to survive unchanged in a final theory is quantum mechanics. This is not only because quantum mechanics is the basis of all of our present understanding of matter and force and has passed extraordinarily stringent experimental tests; more important is the fact that no one has been able to think of any way to change quantum mechanics in any way that would preserve its successes without leading to logical absurdities.

Although quantum mechanics provides the stage on which all natural phenomena are played, in itself it is an empty stage. Quantum mechanics allows us to imagine a huge variety of different possible physical systems: systems composed of any sort of particles interacting through any sorts of forces or even systems that are not composed of particles at all. The history of physics in this century has been marked by the gradual growth of a realization that it is principles of symmetry that dictate the dramatis personae of the drama we observe on the quantum stage. Our present standard model of weak, electromagnetic, and strong forces is based on symmetries: the space-time symmetries of special relativity that require the standard model to be formulated as a theory of fields, and internal symmetries that dictate the existence of the electromagnetic field and the other fields that carry the forces of the standard model. Gravitation, too, can be understood on the basis of a principle of symmetry, the symmetry in Einstein's general theory of relativity that decrees that the laws of nature must not change under any possible changes in how we describe positions in space and time.

Based on this century of experience, it is generally supposed that a final theory will rest on principles of symmetry. We expect that these symmetries will unify gravitation with the weak, electromagnetic, and strong forces of the standard model. But for decades we did not know what these symmetries are, and we had no mathematically satisfactory quantum theory of gravitation that incorporates the symmetry underlying general relativity.

This may now have changed. The past decade has seen the development of a radically new framework for a quantum theory of gravitation and possibly everything else—the theory of strings. String theory has provided our first plausible candidate for a final theory.

The roots of this theory go back to 1968, when elementary

particle theorists were trying to understand the strong nuclear forces without relying on the quantum theory of fields, which was at that time at a low point of popularity. A young theorist at CERN, Gabriel Veneziano, had the idea of simply guessing a formula that would give the probabilities for the scattering of two particles at different energies and angles, and that would have some general properties required by the principles of relativity and quantum mechanics. Using familiar mathematical tools that are learned at one age or another by every physics student, he was able to construct an amazingly simple formula that satisfied all of these conditions. The Veneziano formula attracted a great deal of attention; it was soon generalized by several theorists to other processes and made the basis of a systematic approximation scheme. No one at this time had any idea of any possible application to a quantum theory of gravitation; this work was motivated entirely by the hope of understanding the strong nuclear forces. (The true theory of strong forces, the quantum field theory known as quantum chromodynamics, was then several years in the future.)

In the course of this work it was realized that Veneziano's formula and its extensions and generalizations were not merely lucky guesses but the theory of a new kind of physical entity, a relativistic quantum-mechanical *string*. Of course, ordinary strings are composed of particles like protons and neutrons and electrons, but these new strings are different; *they* are the things of which the protons and neutrons are supposed to be composed. It is not that someone suddenly had an inspiration that matter is composed of strings and then went on to develop a theory based on this idea; the theory of strings had been discovered before anyone realized that it *was* a theory of strings.

These strings can be visualized as tiny one-dimensional rips in the smooth fabric of space. Strings can be open, with two free ends, or closed, like a rubber band. As they fly around in space, the strings vibrate. Each string can be found in any one of an infinite number of possible states (or *modes*) of vibration, much like the various overtones produced by a vibrating tuning fork or violin string. The vibrations of ordinary violin strings die down with time because the energy of vibration of a violin string tends to be converted into a random motion of the atoms of which the violin string is composed, a motion we observe as heat. In contrast, the strings that concern us here are truly fundamental and keep vibrating forever; they are not composed of atoms or anything else, and there is no place for their energy of vibration to go.

The strings are supposed to be very small, so, when a string is observed without probing to very short distances, it appears like a point particle. Because the string can be in any one of an infinite number of possible modes of vibration, it appears like a particle that can belong to any one of an infinite number of possible species, the species corresponding to the mode in which the string is vibrating.

The early versions of string theory were not without problems. Calculations showed that among the infinite number of modes of vibration of a closed string there was one mode in which the string would appear like a particle with zero mass and a spin twice that of the photon. Remember, the discovery of string theories stemmed from Veneziano's efforts to understand the strong nuclear forces, and these string theories were originally conceived to be theories of the strong forces and the particles on which they act. No particle that feels the effects of the strong nuclear forces is known with this mass and spin, and we expect that, if any such particle did exist, it would have been discovered long ago, so this was a serious conflict with experiment.

But there *does* exist a particle with zero mass and a spin twice that of the photon. It is not a particle that feels the strong nuclear forces; it is the graviton, the particle of gravitational radiation. Furthermore, it has been known since the 1960s that any theory of a particle with this spin and mass would have to look more or less the same as general relativity. The massless particle that had been found theoretically in the early days of string theory differed from the true graviton in only one important respect: the exchange of this new massless particle would produce forces that are like gravitational forces but a hundred trillion trillion trillion times stronger.

As often happens in physics, the string theorists had discovered the right solution to the wrong problem. The idea gradually gained ground in the early 1980s that the new massless particle that had been found as a mathematical consequence of string theories was not some sort of strongly interacting analogue to the graviton—it was in fact the true graviton. In order to give gravitational forces the correct strength, it was necessary to increase the string tension in the basic equations of string theory so much that the difference in energy between the lowest and next-to-lowest states of a string would be not the piddling few hundred million volts that is characteristic of nuclear phenomena but rather something like the Planck energy, the million million million billion volts energy at which gravitation becomes as strong as all the other forces.* This is so high an energy that all the particles of the standard model—all the quarks and photons and gluons and so on—must be identified with the lowest modes of vibration of the string; otherwise it would take so much energy to produce them that they would never have been discovered.

From this point of view a quantum field theory like the stan-

---

*Recall that a volt when used as a unit of energy is the energy acquired by one electron in being pushed through a wire by a 1-volt electric battery from one pole of the battery to the other.

dard model is a low-energy approximation to a fundamental theory that is not a theory of fields at all, but a theory of strings. We now think that such quantum field theories work as well as they do at the energies accessible in modern accelerators not because nature is ultimately described by a quantum field theory but because *any* theory that satisfies the requirements of quantum mechanics and special relativity looks like a quantum field theory at sufficiently low energy. Increasingly we regard the standard model as an *effective field theory,* the adjective "effective" serving to remind us that such theories are only low-energy approximations to a very different theory, perhaps a string theory. The standard model has been at the core of modern physics, but this shift in attitude toward quantum field theory may mark the beginning of a new, postmodern, era in physics.

Because string theories incorporate gravitons and a host of other particles, they provided for the first time the basis for a possible final theory. Indeed, because a graviton seems to be an unavoidable feature of any string theory, one can say that string theory explains why gravitation exists. Edward Witten, who later became a leading string theorist, had learned of this aspect of string theories in 1982 from a review article by the CalTech theorist John Schwarz and called this insight "the greatest intellectual thrill of my life."

String theories also seem to have solved the problem of infinities that had plagued all earlier quantum theories of gravitation. Although a string may look like a point particle, the most important thing about them is that they are *not* points but extended objects. The infinities in ordinary quantum field theories can be traced to the fact that the fields describe point particles. (E.g., the inverse-square law gives an infinite force when we put two point electrons at the same position.) On the other hand, properly formulated string theories seem to be free of any infinities.

Interest in string theories really began to take off in 1984, when Schwarz, together with Michael Green of Queen Mary College, London, showed that two specific string theories passed a test for mathematical consistency that had been failed by previously studied string theories. The most exciting feature of the work of Green and Schwarz was its suggestion that string theories have the kind of rigidity that we look for in a really fundamental theory—although one can conceive of a vast number of different open string theories, it seemed that only two of them made mathematical sense. The enthusiasm over string theories reached fever pitch when one team of theorists showed that the low-energy limit of one of these two Green-Schwarz theories had a striking resemblance to our present standard model of weak, strong, and electromagnetic forces, and another team (the "Princeton String Quartet") found a few more string theories with an even closer match to the standard model. Many theorists began to suspect that a final theory was at hand.

Since then enthusiasm has cooled somewhat. It is now understood that there are thousands of string theories that are mathematically consistent in the same way as the two Green-Schwarz theories. All these theories satisfy the same underlying symmetry, known as *conformal symmetry,* but this symmetry is not taken from observation of nature, like Einstein's principle of relativity; rather, conformal symmetry seems to be necessary in order to guarantee the quantum-mechanical consistency of the theories. From this point of view, the thousands of individual string theories merely represent different ways of satisfying the demands of conformal symmetry. It is widely believed that these different string theories are not really different theories, but rather represent different ways of solving the same underlying theory. But we are not certain of this, and no one knows what that underlying theory might be.

Each of the thousands of individual string "theories" has its own space-time symmetries. Some satisfy Einstein's principle of relativity; others do not even have anything that we would recognize as ordinary three-dimensional space. Each string theory also has its own internal symmetries, of the same general sort as the internal symmetries that underlie our present standard model of weak, electromagnetic, and strong forces. But a major difference between the string theories and all earlier theories is that the space-time and internal symmetries are not put in by hand; they are mathematical consequences of the particular way that the rules of quantum mechanics (and the conformal symmetry thus required) are satisfied in each particular string theory. String theories therefore potentially represent a major step toward a rational explanation of nature. They may also be the richest mathematically consistent theories compatible with the principles of quantum mechanics and in particular the only such theories that incorporate anything like gravitation.

A fair fraction of today's young theoretical physicists are working on string theory. Some encouraging results have emerged. For instance, it is natural in string theories for the intrinsic strengths of the strong and electroweak forces to become equal at a very high energy, related to the string tension, even though there is no separate symmetry that unifies these forces. But so far no detailed quantitative predictions have emerged that would allow a decisive test of string theory.

This impasse has led to an unfortunate split in the community of physicists. String theory is very demanding; few of the theorists who work on other problems have the background to understand technical articles on string theory, and few of the string theorists have time to keep up with anything else in physics, least of all with high-energy experiments. Some of my colleagues have reacted to this unhappy predicament with some hostility to string theory. I do not share this feeling. String theory provides our only present source of candidates for a final theory—how could anyone expect that many of the brightest young theorists would *not* work on it? It is a pity that it has not yet been more successful, but string theorists like everyone else are trying to make the best of a very difficult moment in the history of physics. We simply have to hope either that string theory will become more successful or that new experiments will open up progress in other directions.

Unfortunately no one has yet found a specific string theory that exactly matches the particular space-time and internal symmetries and the menu of quarks and leptons that we see in nature. Furthermore, we do not yet know how to enumerate the possible string theories or to evaluate their properties. To solve these problems, it seems to be necessary to invent new methods of calculation that go beyond the techniques that have worked so well in the past. In quantum electrodynamics, for instance, we can calculate the effect of exchanging two photons between electrons in an atom as a small correction to the effect of exchanging one photon, and then can calculate the effect of exchanging three photons as an even smaller correction, and so on, stopping the chain of calculations whenever the remaining corrections are too small to be of interest. This method of calculation is known as perturbation theory. But the crucial questions of string theory involve the exchange of infinite numbers of strings, and so cannot be addressed by perturbation theory.

Matters are even worse than this. Even if we did know how to deal mathematically with these string theories, and even if we could identify one of them that corresponds to what we see in nature, at present we have no criterion that would allow us to tell *why* that string theory is the one that applies to the real world. Once again I repeat: the aim of physics at its most fundamental level is not just to describe the world but to explain why it is the way it is.

In searching for a criterion that would allow us to choose the true string theory, we may be forced to invoke a principle

with a dubious status in physics, known as the *anthropic principle,* which states that the laws of nature should allow the existence of intelligent beings that can ask about the laws of nature.

The idea of an anthropic principle began with the remark that the laws of nature seem surprisingly well suited to the existence of life. A famous example is provided by the synthesis of the elements. According to modern ideas, this synthesis began when the universe was about three minutes old (before then it was too hot for protons and neutrons to stick together in atomic nuclei) and was later continued in stars. It had originally been thought that the elements were formed by adding one nuclear particle at a time to atomic nuclei, starting with the simplest element, hydrogen, whose nucleus consists of just one particle (a proton). But, although there was no trouble in building up helium nuclei, which contain four nuclear particles (two protons and two neutrons), there is no stable nucleus with five nuclear particles and hence no way to take the next step. The solution found eventually by Edwin Salpeter in 1952 is that two helium nuclei can come together in stars to form the unstable nucleus of the isotope beryllium 8, which occasionally before it has a chance to fission into two helium nuclei absorbs yet another helium nucleus and forms a nucleus of carbon. However, as emphasized in 1954 by Fred Hoyle, in order for this process to account for the observed cosmic abundance of carbon, there must be a state of the carbon nucleus that has an energy that gives it an anomalously large probability of being formed in the collision of a helium nucleus and a nucleus of beryllium 8. (Precisely such a state was subsequently found by experimenters working with Hoyle.) Once carbon is formed in stars, there is no obstacle to building up all the heavier elements, including those like oxygen and nitrogen that are necessary for known forms of life. But in order for this to work, the energy of this state of the carbon nucleus must be very close to the energy of a nucleus of beryllium 8 plus the energy of a helium nucleus. If the energy of this state of the carbon nucleus were too large or too small, then little carbon or heavier elements would be formed in stars, and with only hydrogen and helium there would be no way that life could arise. The energies of nuclear states depend in a complicated way on all the constants of physics, such as the masses and electric charges of the different types of elementary particles. It seems at first sight remarkable that these constants should take just the values that are needed to make it possible for carbon to be formed in this way.

The evidence that the laws of nature have been fine-tuned to make life possible does not seem to me very convincing. For one thing, a group of physicists has recently shown that the energy of the state of carbon in question could be increased appreciably without significantly reducing the amount of carbon produced in stars. Also, if we change the constants of nature we may find many other states of the carbon nucleus and other nuclei that might provide alternative pathways for the synthesis of elements heavier than helium. We do not have any good way of estimating how improbable it is that the constants of nature should take values that are favorable for intelligent life.

Whether or not the anthropic principle is needed to explain anything like the energies of nuclear states, there is a context in which it would be only common sense. Perhaps the different logically acceptable universes all in some sense exist, each with its own set of fundamental laws. If this is true then there are certainly many universes whose laws or history make them inhospitable to intelligent life. But any scientist who asks why the world is the way it is must be living in one of the other universes, in which intelligent life *could* arise.*

---

*A Soviet emigré physicist told me that a few years ago a joke was circulating in Moscow, to the effect that the anthropic principle explains why life is so

The weak point in this interpretation of the anthropic principle is that the meaning of a multiplicity of universes is not at all clear. One very simple possibility proposed by Hoyle is that the constants of nature vary from region to region, so that each region of the universe is a sort of subuniverse. The same sort of interpretation of multiple universes might be possible if what we usually call the constants of nature were different in different epochs of the history of the universe. Lately there has been much discussion of a more revolutionary possibility, that our universe and the other logically possible universes with other final laws are in some way spun off from a grander mega-universe. For instance, in recent attempts to apply quantum mechanics to gravitation, it is observed that, although ordinary empty space seems placid and featureless like the surface of the ocean seen from high altitudes, when viewed very closely space seethes with quantum fluctuations, so much so that "wormholes" can open up that connect parts of the universe with other parts that are distant in space and time. In 1987 (following earlier work by Stephen Hawking, James Hartle, and others), Sidney Coleman at Harvard showed that the effect of a wormhole opening or closing is just to change the various constants appearing in the equations that govern various fields. Just as in the many-worlds interpretation of quantum mechanics, the wave function of the universe breaks up into a vast number of terms, in each of which the "constants" of nature take different values, with various different probabilities. In any theories of these various types, it is only common sense that we would find ourselves in a region of space or in an epoch of cosmic history or in a term in the wave function in which the "constants" of nature

---

miserable. There are many more ways for life to be miserable than happy; the anthropic principle only requires that the laws of nature should allow the existence of intelligent beings, not that these beings should enjoy themselves.

happen to take values favorable to the existence of intelligent life.

Physicists will certainly keep trying to explain the constants of nature without resort to anthropic arguments. My own best guess is that we are going to find that in fact all of the constants of nature (with one possible exception) are fixed by symmetry principles of one sort or other and that the existence of some form of life will turn out not to require any very impressive fine-tuning of the laws of nature. The one constant of nature that may have to be explained by some sort of anthropic principle is the one known as the *cosmological constant*.

The cosmological constant originally appeared in physical theory in Einstein's first attempt to apply his new general theory of relativity to the whole universe. In this work he assumed as was usual at that time that the universe is static, but he soon found that his gravitational field equations in their original form when applied to the whole universe did not have any static solutions. (This conclusion actually has nothing specifically to do with general relativity; in Newton's theory of gravitation also we might find solutions with galaxies rushing toward each other under the influence of their mutual gravitation, and other solutions with galaxies rushing apart from each other as a consequence of an initial explosion, but we would not expect that the average galaxy could be just hanging more or less at rest in space.) In order to allow a static universe, Einstein decided to change his theory. He introduced a term in his equations that would produce something like a repulsive force at large distances and that could thus balance the attractive force of gravitation. This term involves one free constant, which in Einstein's static cosmology determined the size of the universe, and thus became known as the cosmological constant.

This was in 1917. Because of the war, Einstein did not know that an American astronomer, Vesto Slipher, had already found indications that the galaxies (as we now call them) are rushing apart, so that the universe is in fact not static but expanding. The expansion was confirmed and its rate measured after the war by Edwin Hubble, using the new 100-inch telescope at Mount Wilson. Einstein came to regret mutilating his equations by introducing the cosmological constant. However the possibility of a cosmological constant did not go away so easily.

For one thing, there is no reason *not* to include a cosmological constant in the Einstein field equations. Einstein's theory had been based on a symmetry principle, which states that the laws of nature should not depend on the frame of reference in space and time that we use to study these laws. But his original theory was not the most general theory allowed by that symmetry principle. There is a vast number of possible allowed terms that might be added to the field equations, whose effects would be negligible over astronomical distances and can therefore safely be ignored. Aside from these, there is only one term that could be added to the Einstein field equations without violating the fundamental symmetry principle of general relativity and that would be important in astronomy: it is the term involving the cosmological constant. Einstein in 1915 operated under the assumption that the field equations should be chosen to be as simple as possible. The experience of the past three-quarters of a century has taught us to distrust such assumptions; we generally find that any complication in our theories that is not forbidden by some symmetry or other fundamental principle actually occurs. It is thus not enough to say that a cosmological constant is an unnecessary complication. Simplicity, like everything else, must be explained.

In quantum mechanics the problem is worse. The various fields that inhabit our universe are subject to continual quantum fluctuations that give an energy even to nominally empty space. This energy is observable only through its gravitational effects; energy of any sort generates gravitational fields and is in turn acted on by gravitational fields, so an energy filling all space could have important effects on the expansion of the universe. We cannot actually calculate the energy per volume that is produced by these quantum fluctuations; using the simplest approximations, it turns out to be infinite. But, with any reasonable guess at how to throw away the high-frequency fluctuations that are responsible for the infinity, the vacuum energy per volume comes out to be enormously large: it is about a trillion trillion trillion trillion trillion trillion trillion trillion trillion trillion times larger than is allowed by the observed rate of expansion of the universe. This must be the worst failure of an order-of-magnitude estimate in the history of science.

If this energy of empty space is positive, then it produces a gravitational repulsion between particles of matter at very large distances, precisely like the term involving the cosmological constant that Einstein added to his field equations in 1917. We can therefore regard the energy due to quantum fluctuations as merely making a contribution to a "total" cosmological constant; the expansion of the universe is affected only by this total cosmological constant, not by the cosmological constant in the field equations of general relativity or by the quantum vacuum energy separately. This opens up the possibility that the problem of the cosmological constant and the problem of the energy of empty space may cancel. In other words, there may be a *negative* cosmological constant in the Einstein field equations that merely cancels the effect of the enormous vacuum energy owing to quantum fluctuations. But in order to be consistent with what we know about the expansion of the universe, the total cosmological constant must be so small that these two terms in the total cosmological constant would have to cancel to 120 decimal places. This is not the sort of thing that we would be happy to leave unexplained.

Theoretical physicists have been trying for years to understand the cancellation of the total cosmological constant, so far without having found any convincing explanation. String theory, if anything, makes the problem worse. The many different string theories each yield a different value for the total cosmological constant (including the effects of vacuum quantum fluctuations), but in the general case this comes out enormously too large. With the total cosmological constant this large, space would be so radically curved that it would bear no resemblance to the familiar three-dimensional space of Euclidean geometry in which we live.

If all else fails, we may be thrown back on an anthropic explanation. There may in some sense or other be many different "universes," each with its own value for the cosmological constant. If this were true, then the only universe in which we could expect to find ourselves is one in which the total cosmological constant is small enough to allow life to arise and evolve. To be specific, if the total cosmological constant were large and negative, the universe would run through its life cycle of expansion and contraction too rapidly for life to have time to appear. On the other hand, if the total cosmological constant were large and positive, the universe would expand forever, but the repulsive force produced by the cosmological constant would prevent the gravitational clumping together of matter to form galaxies and stars in the early universe and therefore give life no place to appear. Perhaps the true string theory is the one (if there is only one) that leads to a total cosmological constant in the relatively narrow range of small values for which life could appear.

One of the intriguing consequences of this line of thought is that there is no reason why the total cosmological constant (including the effects of vacuum quantum fluctuations) should be strictly zero; the anthropic principle requires only that it be small enough to allow galaxies to form and to survive for billions of years. In fact, for some time there have been hints from astronomical observation that the total cosmological constant is not zero, but small and positive.

One of these hints is provided by the celebrated "cosmological missing-mass" problem. The most natural value for the mass density of the universe (and the value required by currently popular cosmological theories) is that density whose gravitational attraction would just barely allow the universe to keep expanding forever. But this density is about five to ten times larger than is contributed by the mass in clusters of galaxies (as inferred from studies of the motions of the galaxies in these clusters). The missing mass might well be dark matter of some sort, but there is another possibility. As already mentioned the effect of a positive cosmological constant is precisely like that of a uniform constant energy density, which according to Einstein's famous relation between energy and mass is equivalent to a uniform constant mass density. It is thus possible that the missing 80%–90% of the cosmic "mass" density is provided by a positive total cosmological constant rather than by any sort of real matter.

This is not to say that there is no difference between a real matter density and a positive total cosmological constant. The universe is expanding, so whatever the density of real matter is now, it was much larger in the past. In contrast, the total cosmological constant is constant in time, and so is the matter density to which it is equivalent. The higher the matter density, the faster the expansion of the universe, so the expansion rate in the past would have been much greater if the missing "mass" is ordinary matter rather than an effect of a cosmological constant.

Another hint that points more specifically toward a positive total cosmological constant comes from a long-standing problem with the age of the universe. In conventional cosmological theories we can use the observed expansion rate of the universe to infer that the universe is about seven to twelve billion years old. But the ages of the globular clusters of stars within our own galaxy are usually estimated to be about twelve to fifteen billion years. We are confronted with the prospect of a universe younger than the globular clusters it contains. To avoid this paradox we would have to adopt the lowest estimated ages for the globular clusters and the highest estimates for the age of the universe. On the other hand, as we have seen, the introduction of a positive cosmological constant in place of dark matter would have the effect of decreasing our estimate of the rate of expansion of the universe in the past and hence of increasing the age of the universe that we would infer from any given present rate of expansion. For instance, if the cosmological constant contributes 90% of the cosmic "mass" density, then even for the highest estimates of the present rate of expansion of the universe the age of the universe would be eleven billion years rather than only seven billion years, so that any serious discrepancy with the globular cluster ages would be removed.

A positive cosmological constant that provides 80%–90% of the present cosmic "mass" density is well within the limits that would allow for the existence of life. We know that quasars and presumably also galaxies were already condensing out of the big bang so early that the universe was then only one-sixth as large as it is now, because we see light from quasars whose wavelength has increased (i.e., redshifted) sixfold. At that time the real mass density of the universe was six cubed or about two hundred times larger than it is now, so a cosmological constant corresponding to a mass density that is only five to ten times larger than the *present* mass density could have had no significant effect on the formation of galaxies *then,* though it would have prevented a more recent formation of galaxies. A cosmological constant that provides a "mass" density five to ten times larger than the present cosmic density of matter is thus very roughly what we should expect on anthropic grounds.

Fortunately this is a question that (unlike many of the others discussed in this chapter) may be settled before long by astronomical observation. As we have seen, the expansion rate of the universe in the past would have been much greater if the missing mass is made up of ordinary matter rather than being due to a cosmological constant. This difference in expansion rates affects the geometry of the universe and the paths of light rays in ways that could be detected by astronomers. (E.g., it would alter the numbers of galaxies observed to be rushing away from us at various velocities, as well as the numbers of galactic gravitational lenses—galaxies whose gravitational field bends the light of more distant objects enough to form multiple images.) The observations so far are inconclusive, but these issues are being actively pursued at several observatories, and should eventually confirm or rule out a cosmological constant that provides 80%–90% of the present "mass" density of the universe. Such a cosmological constant is so much less than would have been expected from estimates of quantum fluctuations that it would be difficult to understand on any other than anthropic grounds. Thus, if such a cosmological constant is confirmed by observation, it will be reasonable to infer that our own existence plays an important part in explaining why the universe is the way it is.

For what it is worth, I hope that this is not the case. As a theoretical physicist, I would like to see us able to make precise predictions, not vague statements that certain constants have to be in a range that is more or less favorable to life. I hope that string theory really will provide a basis for a final theory and that this theory will turn out to have enough predictive power to be able to prescribe values for all the constants of nature, including the cosmological constant. We shall see.

CHAPTER X

FACING FINALITY

*The Pole at last! The prize of three centuries. . . . I can-
not bring myself to realize it. It seems all so simple and
commonplace.*

Robert Peary, diary, quoted by him in *The North Pole*

I t is difficult to imagine that we could ever be in possession of
final physical principles that have no explanation in terms of
deeper principles. Many people take it for granted that instead
we shall find an endless chain of deeper and deeper principles.
For example, Karl Popper, the dean of modern philosophers
of science, rejects "the idea of an ultimate explanation." He
maintains that "every explanation may be further explained,
by a theory or conjecture of a higher degree of universality.
There can be no explanation which is not in need of a further
explanation. . . ."

Popper and the many others who believe in an infinite chain
of more and more fundamental principles might turn out to be
right. But I do not think that this position can be argued on the

grounds that no one has yet found a final theory. That would be like a nineteenth-century explorer arguing that, because all previous arctic explorations over hundreds of years had always found that however far north they penetrated there was still more sea and ice left unexplored to the north, either there was no North Pole or in any case no one would ever reach it. Some searches do come to an end.

There seems to be a widespread impression that scientists in the past have often deluded themselves that they had found a final theory. They are imagined to be like the explorer Frederick Cook in 1908, who only thought that he had reached the North Pole. It is fancied that scientists are given to constructing elaborate theoretical schemes that they declare to be the final theory, which they then doggedly defend until overwhelming experimental evidence reveals to new generations of scientists that these schemes are all wrong. But, as far as I know, no reputable physicist in this century has claimed that a final theory had actually been found. Physicists do sometimes underestimate the distance that still must be traveled before a final theory is reached. Recall Michelson's 1902 prediction that "the day appears not far distant when the converging lines from many apparently remote regions of thought will meet on . . . common ground." More recently Stephen Hawking in assuming the Lucasian Chair of Mathematics at Cambridge (the chair held previously by Newton and Dirac) suggested in his inaugural lecture that the "extended supergravity" theories then fashionable were going to provide a basis for something like a final theory. I doubt that Hawking would suggest that today. But neither Michelson nor Hawking ever claimed that a final theory had already been achieved.

If history is any guide at all, it seems to me to suggest that there *is* a final theory. In this century we have seen a convergence of the arrows of explanation, like the convergence of meridians toward the North Pole. Our deepest principles, although not yet final, have become steadily more simple and economical. We saw this convergence here in explaining the properties of a piece of chalk, and I have observed it within the span of my own career in physics. When I was a graduate student, I had to learn a vast amount of miscellaneous information about the weak and strong interactions of the elementary particles. Today, students of elementary particle physics learn the standard model and a great deal of mathematics and often little else. (Professors of physics sometimes wring their hands over how little of the actual phenomena of elementary particle physics the students know, but I suppose that those who taught me at Cornell and Princeton were wringing their hands over how little atomic spectroscopy *I* knew.) It is very difficult to conceive of a regression of more and more fundamental theories becoming steadily simpler and more unified, without the arrows of explanation having to converge somewhere.

It is conceivable but unlikely that the chains of more and more fundamental theories neither go on forever nor come to an end. The Cambridge philosopher Michael Redhead suggests that they may turn back on themselves. He notes that the orthodox Copenhagen interpretation of quantum mechanics requires the existence of a macroscopic world of observers and measuring apparatuses, which in turn is explained in terms of quantum mechanics. This view seems to me to provide one more example of what is wrong with the Copenhagen interpretation of quantum mechanics, and with the difference between the ways that it treats quantum phenomena and the observers that study them. In the realist approach to quantum mechanics of Hugh Everett and others, there is just one wave function describing all phenomena, including experiments and observers, and the fundamental laws are those that describe the evolution of this wave function.

Yet more radical is the suggestion that at bottom we shall find that there is no law at all. My friend and teacher John Wheeler has occasionally suggested that there is no fundamental law and that all the laws we study today are imposed on nature by the way that we make observations. Along somewhat different lines, the Copenhagen theorist Holger Nielsen has proposed a "random dynamics," according to which, whatever we assume about nature at very short distances or very high energies, the phenomena accessible in our laboratories will look about the same.

Both Wheeler and Nielsen simply seem to me to be merely pushing back the problem of the final laws. Wheeler's world without law still needs metalaws to tell us how our observations impose regularities on nature, among which metalaws is quantum mechanics itself. Nielsen likewise needs some sort of metalaw to explain how the appearance of nature changes as we change the scale of distances and energies at which we make our measurements, and for this purpose assumes the validity of what are called renormalization group equations, whose origin in a world without law is certainly problematic. I expect that all attempts to do without fundamental laws of nature, if successful at all, simply result in the introduction of metalaws that describe how what we *now* call laws came about.

There is another possibility that seems to me more likely and much more disturbing. Perhaps there is a final theory, a simple set of principles from which flow all arrows of explanation, but we shall never learn what it is. For instance, it may be that humans are simply not intelligent enough to discover or to understand the final theory. It is possible to train dogs to do all sorts of clever things, but I doubt that anyone will ever train a dog to use quantum mechanics to calculate atomic energy levels. The best reason for hope that our species is intellectually capable of continued future progress is our wonderful ability to link our brains through language, but this may not be enough. Eugene Wigner has warned that "we have no right to expect that our intellect can formulate perfect concepts for the full understanding of inanimate nature's phenomena." So far, fortunately, we do not seem to be coming to the end of our intellectual resources. In physics at any rate each new generation of graduate students seems brighter than the last.

A far more pressing worry is that the effort to discover the final laws may be stopped for want of money. We have a foretaste of this problem in the recent debate in the United States over whether to complete the Super Collider. Its 8-billion-dollar cost spread over a decade is certainly well within our country's capabilities, but even high-energy physicists would hesitate to propose a much more expensive future accelerator.

Beyond the questions about the standard model that we expect to be answered by the Super Collider, there is a level of deeper questions having to do with the unification of strong, electroweak, and gravitational interactions, questions that cannot be directly addressed by any accelerator now conceivable. The really fundamental Planck energy where all these questions could be explored experimentally is about a hundred trillion times higher than the energy that would be available at the Superconducting Super Collider. It is at the Planck energy where all the forces of nature are expected to become unified. Also, this is roughly the energy that according to modern string theories is needed to excite the first modes of vibration of strings, beyond the lowest modes that we observe as ordinary quarks and photons and the other particles of the standard model. Unfortunately, such energies seem hopelessly beyond our reach. Even if the entire economic resources of the human race were devoted to the task, we would not today know how to build a machine that could accelerate particles to such energies. It is not that the energy itself is unavailable—the Planck energy is roughly the same as the chemical energy in a full automobile gasoline tank. The difficult problem is to concentrate all that energy on a single proton or electron. We may learn how to build such accelerators in very different ways from those that are used today, perhaps by using ionized gases to help transfer energy from powerful laser beams to individual charged particles, but even so the reaction rate of particles at this energy would be so small that experiments might be impossible. It is more likely that breakthroughs in theory or in other sorts of experiments will some day remove the necessity of building accelerators of higher and higher energies.

My own guess is that there is a final theory, and we are capable of discovering it. It may be that experiments at the Super Collider will yield such illuminating new information that theorists will be able to complete the final theory without having to study particles at the Planck energy. We may even be able to find a candidate for such a final theory among today's string theories.

How strange it would be if the final theory were to be discovered in our own lifetimes! The discovery of the final laws of nature will mark a discontinuity in human intellectual history, the sharpest that has occurred since the beginning of modern science in the seventeenth century. Can we now imagine what that would be like?

Although it is not hard to conceive of a final theory that does not *have* an explanation in terms of deeper principles, it is very difficult to imagine a final theory that does not *need* such an explanation. Whatever the final theory may be, it will certainly not be *logically* inevitable. Even if the final theory turns out to be a theory of strings that can be expressed in a few simple equations, and even if we can show that this is the only possible quantum-mechanical theory that can describe gravitation along with other forces without mathematical inconsistencies, we will still have to ask why there should be such a thing as gravitation and why nature should obey the rules of quantum mechanics. Why does the universe not consist merely of point particles orbiting endlessly according to the rules of Newtonian mechanics? Why is there anything at all? Redhead probably represents a majority view, in denying that "the aim of some self-vindicating a priori foundation for science is a credible one."

On the other hand, Wheeler once remarked that, when we come to the final laws of nature, we will wonder why they were not obvious from the beginning. I suspect that Wheeler may be correct, but only because by then we will have been trained by centuries of scientific failures and successes to find these laws obvious. Even so, in however attenuated a form, I think the old question, Why? will still be with us. The Harvard philosopher Robert Nozick has grappled with this problem and suggests that instead of trying to deduce the final theory on the basis of pure logic, we should search instead for arguments that would make it somehow more satisfying than a mere brute fact.

In my view, our best hope along this line is to show that the final theory, though not logically inevitable, is logically *isolated*. That is, it may turn out that, although we shall always be able to imagine other theories that are totally different from the true final theory (like the boring world of particles governed by Newtonian mechanics), the final theory we discover is so rigid that there is no way to modify it by a small amount without the theory leading to logical absurdities. In a logically isolated theory every constant of nature could be calculated from first principles; a small change in the value of any constant would destroy the consistency of the theory. The final theory would be like a piece of fine porcelain that cannot be warped without shattering. In this case, although we may still not know why the final theory is true, we would know on the basis of pure mathematics and logic why the truth is not slightly different.

This is not just a possibility—we are already well along the road to such a logically isolated theory. The most fundamental known physical principles are the rules of quantum mechanics, which underlie everything else that we know about matter and its interactions. Quantum mechanics is not logically inevitable; there does not seem to be anything logically impossible about its predecessor, the mechanics of Newton. Yet the best efforts of physicists have failed to discover any way of changing the rules of quantum mechanics *by a small amount* without incurring logical disasters, such as probabilities that come out to be negative numbers.

But quantum mechanics by itself is not a complete physical theory. It tells us nothing about the particles and forces that may exist. Pick up any textbook on quantum mechanics; you find as illustrative examples a weird variety of hypothetical particles and forces, most of which resemble nothing that exists in the real world, but all of which are perfectly consistent with the principles of quantum mechanics and can be used to give students practice in applying these principles. The variety of possible theories becomes much smaller if we consider only quantum-mechanical theories consistent with the special theory of relativity. Most of these theories can be logically ruled out because they would entail nonsense like infinite energies or infinite reaction rates. Even so there are still plenty of logically possible theories, such as the theory of strong nuclear forces known as quantum chromodynamics, with nothing in the universe but quarks and gluons. But most of these theories are ruled out if we also insist that they involve gravitation. It is possible that we will be able to show mathematically that these requirements leave only one logically possible quantum-mechanical theory, perhaps a unique theory of strings. If this is so, then, although there would still be a vast number of other logically possible final theories, there would be only one that describes anything even remotely like our own world.

But why should the final theory describe anything like our world? The explanation might be found in what Nozick has called the *principle of fecundity*. It states that the different logically acceptable universes all in some sense exist, each with its own set of fundamental laws. The principle of fecundity is not itself explained by anything, but at least it has a certain pleasing self-consistency; as Nozick says, the principle of fecundity states "that all possibilities are realized, while it itself is one of those possibilities."

If this principle is true then our own quantum-mechanical world exists, but so does the Newtonian world of particles orbiting endlessly and so do worlds that contain nothing at all and so do countless other worlds that we cannot even imagine. It is not just a matter of the so-called constants of nature varying from one part of the universe to another or from one epoch to another or from one term in the wave function to another. As we have seen, these are all possibilities that might be realized as consequences of some really fundamental theory like quantum cosmology but that would still leave us with the problem of understanding why that fundamental theory is what it is. The principle of fecundity instead supposes that there are entirely different universes, subject to entirely different laws. But, if these other universes are totally inaccessible and unknowable, then the statement that they exist would seem to have no consequences, except to avoid the question of why they do not exist. The problem seems to be that we are trying to be logical about a question that is not really susceptible to logical argument: the question of what should or should not engage our sense of wonder.

The principle of fecundity would provide yet another way

of justifying the use of anthropic reasoning to help explain why the final laws of *our* universe are what they are. There would be many conceivable kinds of universe whose laws or history make them inhospitable to intelligent life, but any scientist who asks why the world is the way it is would have to be living in one of the other universes, in which intelligent life *could* arise. In this way we can immediately rule out the universe governed by Newtonian physics (for one thing, there would be no stable atoms in such a world), or the universe containing nothing at all.

As an extreme possibility, it is possible that there is only one logically isolated theory, with *no* undetermined constants, that is consistent with the existence of intelligent beings capable of wondering about the final theory. If this could be shown, then we would be as close as anyone could hope to a satisfactory explanation of why the world is the way it is.

What would be the effect of the discovery of such a final theory? Of course, a definite answer will have to wait until we know the final theory. We may discover things about the governance of the world that are as surprising to us as the rules of Newtonian mechanics would have been to Thales. But about one thing we can be sure: the discovery of a final theory would not end the enterprise of science. Even apart from problems that need to be studied for the purposes of technology or medicine, there would still be plenty of problems of pure science that will be pursued because scientists expect these problems to have beautiful solutions. Right now in physics alone there are phenomena like turbulence and high-temperature superconductivity that are expected to have profound and beautiful explanations. No one knows how galaxies formed or how the genetic mechanism got started or how memories are stored in the brain. None of these problems is likely to be affected by the discovery of a final theory.

On the other hand, the discovery of a final theory may have effects far beyond the borders of science. The minds of many people today are afflicted with various irrational misconceptions, ranging from relatively harmless superstitions like astrology to ideologies of the most vicious sort. The fact that the fundamental laws of nature remain obscure makes it that much easier for people to hope that some day their own favorite irrationalities will find a respectable place within the structure of science. It would be foolish to expect that any discovery of science could in itself purge the human race of all its misconceptions, but the discovery of the final laws of nature will at least leave less room in the imagination for irrational beliefs.

Still, with the discovery of a final theory we may regret that nature has become more ordinary, less full of wonder and mystery. Something like this has happened before. Throughout most of human history our maps of the earth have shown great unknown spaces, that the imagination could fill with dragons and golden cities and anthropophagi. The search for knowledge was largely a matter of geographical exploration. When Tennyson's Ulysses set out to "follow knowledge like a sinking star, beyond the utmost bounds of human thought," he sailed out into the unknown Atlantic, "beyond the sunset, and the baths of all the western stars." But today every acre of the earth's land surface has been mapped, and the dragons are all gone. With the discovery of the final laws, our daydreams will again contract. There will be endless scientific problems and a whole universe left to explore, but I suspect that scientists of the future may envy today's physicists a little, because we are still on our voyage to discover the final laws.

CHAPTER XI

WHAT ABOUT GOD?

"You know," said Port, and his voice sounded unreal, as voices are likely to do after a long pause in an utterly silent spot, "the sky here's very strange. I often have the sensation when I look at it that it's a solid thing up there, protecting us from what's behind."

Kit shuddered slightly as she said: "From what's behind?"

"Yes."

"But what is behind?" Her voice was very small.

"Nothing, I suppose. Just darkness. Absolute night."

Paul Bowles, *The Sheltering Sky*

"The heavens declare the glory of God; and the firmament showeth his handiwork." To King David or whoever else wrote this psalm, the stars must have seemed visible evidence of a more perfect order of existence, quite different from our dull sublunary world of rocks and stones and trees. Since David's day the sun and other stars have lost their special status; we understand that they are spheres of glowing gas, held together by gravitation, and supported against collapse by pressure that is maintained by the heat rising up from thermonuclear reactions in the stars' cores. The stars tell us nothing more or less about the glory of God than do the stones on the ground around us.

If there were anything we could discover in nature that *would* give us some special insight into the handiwork of God, it would have to be the final laws of nature. Knowing these laws, we would have in our possession the book of rules that governs stars and stones and everything else. So it is natural that Stephen Hawking should refer to the laws of nature as "the mind of God." Another physicist, Charles Misner, used similar language in comparing the perspectives of physics and chemistry: "The organic chemist, in answer to the question, 'Why are there ninety-two elements, and when were they produced?' may say 'The man in the next office knows that.' But the physicist, being asked, 'Why is the universe built to follow certain physical laws and not others?' may well reply, 'God knows.'" Einstein once remarked to his assistant Ernst Straus that "What really interests me is whether God had any choice in the creation of the world." On another occasion he described the aim of the enterprise of physics as "not only to know how nature is and how her transactions are carried through, but also to reach as far as possible the utopian and seemingly arrogant aim of knowing why *nature is thus and not otherwise*. . . . Thereby one experiences, so to speak, that God Himself could not have arranged these connections in any other way than that which factually exists. . . . This is the Promethean element of the scientific experience. . . . Here has always been for me the particular magic of scientific effort." Einstein's religion was so vague that I suspect that he meant this metaphorically, as suggested by his "so to speak." It is doubtless because physics is so fundamental that this metaphor is natural to physicists. The theologian Paul Tillich once observed that among scientists only physicists seem capable of using the word "God" without embarrassment. Whatever one's religion or lack of it, it is an irresistible metaphor to speak of the final laws of nature in terms of the mind of God.

I encountered this connection once in an odd place, in the Rayburn House Office Building in Washington. When I testified there in 1987 in favor of the Superconducting Super Collider (SSC) project before the House Committee on Science, Space, and Technology, I described how in our study of elementary particles we are discovering laws that are becoming increasingly coherent and universal, and how we are beginning to suspect that this is not merely an accident, that there is a beauty in these laws that mirrors something that is built into the structure of the universe at a very deep level. After I made these remarks there were remarks by other witnesses and questions from members of the committee. There then ensued a dialogue between two committee members, Representative Harris W. Fawell, Republican of Illinois, who had generally been favorable to the Super Collider project, and Representative Don Ritter, Republican of Pennsylvania, a former metallurgical engineer who is one of the most formidable opponents of the project in Congress:

MR. FAWELL: . . . Thank you very much. I appreciate the testimony of all of you. I think it was excellent. If ever I would want to explain to one and all the reasons why the SSC is needed I am sure I can go to your testimony. It would be very helpful. I wish sometimes that we have some one word that could say it all and that is kind of impossible. I guess perhaps Dr. Weinberg you came a little close to it and I'm not sure but I took this down. You said you suspect that it isn't all an accident that there are rules which govern matter and I jotted down, will this make us find God? I'm sure you didn't make that claim, but it certainly will enable us to understand so much more about the universe?

MR. RITTER: Will the gentleman yield on that? If the gentleman would yield for a moment I would say . . .

MR. FAWELL: I'm not sure I want to.

MR. RITTER: If this machine does that I am going to come around and support it.

I had enough sense to stay out of this exchange, because I did not think that the congressmen wanted to know what I thought about finding God at the SSC and also because it did not seem to me that letting them know what I thought about this would be helpful to the project.

Some people have views of God that are so broad and flexible that it is inevitable that they will find God wherever they look for Him. One hears it said that "God is the ultimate" or "God is our better nature" or "God is the universe." Of course, like any other word, the word "God" can be given any meaning we like. If you want to say that "God is energy," then you can find God in a lump of coal. But if words are to have any value to us, we ought to respect the way that they have been used historically, and we ought especially to preserve distinctions that prevent the meanings of words from merging with the meanings of other words.

In this spirit, it seems to me that if the word "God" is to be of any use, it should be taken to mean an interested God, a creator and lawgiver who has established not only the laws of nature and the universe but also standards of good and evil, some personality that is concerned with our actions, something in short that it is appropriate for us to worship.* This is the God that has mattered to men and women throughout history. Scientists and others sometimes use the word "God" to mean something so abstract and unengaged that He is hardly to be

---

*It should be apparent that in discussing these things I am speaking only for myself and that in this chapter I leave behind me any claim to special expertise.

distinguished from the laws of nature. Einstein once said that he believed in "Spinoza's God who reveals Himself in the orderly harmony of what exists, not in a God who concerns himself with fates and actions of human beings." But what possible difference does it make to anyone if we use the word "God" in place of "order" or "harmony," except perhaps to avoid the accusation of having no God? Of course, anyone is free to use the word "God" in that way, but it seems to me that it makes the concept of God not so much wrong as unimportant.

Will we find an interested God in the final laws of nature? There seems something almost absurd in asking this question, not only because we do not yet know the final laws, but much more because it is difficult even to imagine being in the possession of ultimate principles that do not need any explanation in terms of deeper principles. But premature as the question may be, it is hardly possible not to wonder whether we will find any answer to our deepest questions, any sign of the workings of an interested God, in a final theory. I think that we will not.

All our experience throughout the history of science has tended in the opposite direction, toward a chilling impersonality in the laws of nature. The first great step along this path was the demystification of the heavens. Everyone knows the key figures: Copernicus, who proposed that the earth is not at the center of the universe, Galileo, who made it plausible that Copernicus was right, Bruno who guessed that the sun is only one of a vast number of stars, and Newton who showed that the same laws of motion and gravitation apply to the solar system and to bodies on the earth. The key moment I think was Newton's observation that the same law of gravitation governs the motion of the moon around the earth and a falling body on the surface of the earth. In our own century the demystification of the heavens was taken a step farther by the American astronomer Edwin Hubble. By measuring the distance to the Andromeda Nebula, Hubble showed that this and by inference thousands of other similar nebulas were not just outlying parts of our galaxy but galaxies in their own right, quite as impressive as our own. Modern cosmologists even speak of a Copernican principle: the rule that no cosmological theory can be taken seriously that puts our own galaxy at any distinctive place in the universe.

Life, too, has been demystified. Justus von Liebig and other organic chemists in the early nineteenth century demonstrated that there was no barrier to the laboratory synthesis of chemicals like uric acid that are associated with life. Most important of all were Charles Darwin and Alfred Russel Wallace, who showed how the wonderful capabilities of living things could evolve through natural selection with no outside plan or guidance. The process of demystification has accelerated in this century, in the continued success of biochemistry and molecular biology in explaining the workings of living things.

The demystification of life has had a far greater effect on religious sensibilities than has any discovery of physical science. It is not surprising that it is reductionism in biology and the theory of evolution rather than the discoveries of physics and astronomy that continue to evoke the most intransigent opposition.

Even from scientists one hears occasional hints of vitalism, the belief in biological processes that cannot be explained in terms of physics and chemistry. In this century biologists (including antireductionists like Ernst Mayr) have generally steered clear of vitalism, but as late as 1944 Erwin Schrödinger argued in his well-known book *What Is Life?* that "enough is known about the material structure of life to tell exactly why present-day physics cannot account for life." His reason was that the genetic information that governs living organisms is far too stable to fit into the world of continual fluctuations described by quantum mechanics and statistical mechanics. Schrödinger's mistake was pointed out by Max Perutz, the molecular biologist who among other things worked out the structure of hemoglobin: Schrödinger had ignored the stability that can be produced by the chemical process known as enzymatic catalysis.

The most respectable academic critic of evolution may currently be Professor Phillip Johnson of the University of California School of Law. Johnson concedes that evolution has occurred and that it is sometimes due to natural selection, but he argues that there is no "incontrovertible experimental evidence" that evolution is not guided by some divine plan. Of course, one could never hope to prove that no supernatural agency ever tips the scales in favor of some mutations and against others. But much the same could be said of any scientific theory. There is nothing in the successful application of Newton's or Einstein's laws of motion to the solar system that prevents us from supposing that every once in a while some comet gets a small shove from a divine agency. It seems pretty clear that Johnson raises this issue not as a matter of impartial open-mindedness but because for religious reasons he cares very much about life in a way that he does not care about comets. But the only way that any sort of science can proceed is to assume that there is no divine intervention and to see how far one can get with this assumption.

Johnson argues that naturalistic evolution, "evolution that involves no intervention or guidance by a creator outside the world of nature," in fact does not provide a very good explanation for the origin of species. I think he goes wrong here because he has no feeling for the problems that any scientific theory always has in accounting for what we observe. Even apart from outright errors, our calculations and observations are always based on assumptions that go beyond the validity of the theory we are trying to test. There never was a time when the calculations based on Newton's theory of gravitation or any other theory were in perfect agreement with all observations. In the writings of today's paleontologists and evolutionary biologists we can recognize the same state of affairs that is so familiar to us in physics; in using the naturalistic theory of evolution biologists are working with an overwhelmingly successful theory, but one that is not yet finished with its work of explication. It seems to me to be a profoundly important discovery that we can get very far in explaining the world without invoking divine intervention, and in biology as well as in the physical sciences.

In another respect I think that Johnson is right. He argues that there is an incompatibility between the naturalistic theory of evolution and religion as generally understood, and he takes to task the scientists and educators who deny it. He goes on to complain that "naturalistic evolution is consistent with the existence of 'God' only if by that term we mean no more than a first cause which retires from further activity after establishing the laws of nature and setting the natural mechanism in motion."

The inconsistency between the modern theory of evolution and belief in an interested God does not seem to me one of logic—one can imagine that God established the laws of nature and set the mechanism of evolution in motion with the intention that through natural selection you and I would someday appear—but there is a real inconsistency in temperament. After all, religion did not arise in the minds of men and women who speculated about infinitely prescient first causes but in the hearts of those who longed for the continual intervention of an interested God.

The religious conservatives understand, as their liberal opponents seem often not to, how high are the stakes in the debate over teaching evolution in the public schools. In 1983, shortly after coming to Texas, I was invited to testify before a committee of the Texas Senate on a regulation that forbade the teaching of the theory of evolution in state-purchased high-school textbooks unless equal emphasis was given to creationism. One of the members of the committee asked me how the state could support the teaching of a scientific theory like evolution that was so corrosive of religious belief. I replied that just as it would be wrong for those who are emotionally committed to atheism to give evolution more emphasis than would be otherwise appropriate in teaching biology, so it would be inconsistent with the First Amendment to give evolution less emphasis as a means of protecting religious belief. It is simply not the business of the public schools to concern themselves one way or the other with the religious implications of scientific theories. My answer did not satisfy the senator because he knew as I did what would be the effect of a course in biology that gives an appropriate emphasis to the theory of evolution. As I left the committee room, he muttered that "God is still in heaven anyway." Maybe so, but we won that battle; Texas high-school textbooks are now not only allowed but required to teach the modern theory of evolution, and with no nonsense about creationism. But there are many places (today especially in Islamic countries) where this battle is yet to be won and no assurance anywhere that it will stay won.

One often hears that there is no conflict between science and religion. For instance, in a review of Johnson's book, Stephen Gould remarks that science and religion do not come into conflict, because "science treats factual reality, while religion treats human morality." On most things I tend to agree with Gould, but here I think he goes too far; the meaning of religion is defined by what religious people actually believe, and the great majority of the world's religious people would be surprised to learn that religion has nothing to do with factual reality.

But Gould's view is widespread today among scientists and

religious liberals. This seems to me to represent an important retreat of religion from positions it once occupied. Once nature seemed inexplicable without a nymph in every brook and a dryad in every tree. Even as late as the nineteenth century the design of plants and animals was regarded as visible evidence of a creator. There are still countless things in nature that we cannot explain, but we think we know the principles that govern the way they work. Today for real mystery one has to look to cosmology and elementary particle physics. For those who see no conflict between science and religion, the retreat of religion from the ground occupied by science is nearly complete.

Judging from this historical experience, I would guess that, though we shall find beauty in the final laws of nature, we will find no special status for life or intelligence. A fortiori, we will find no standards of value or morality. And so we will find no hint of any God who cares about such things. We may find these things elsewhere, but not in the laws of nature.

I have to admit that sometimes nature seems more beautiful than strictly necessary. Outside the window of my home office there is a hackberry tree, visited frequently by a convocation of politic birds: blue jays, yellow-throated vireos, and, loveliest of all, an occasional red cardinal. Although I understand pretty well how brightly colored feathers evolved out of a competition for mates, it is almost irresistible to imagine that all this beauty was somehow laid on for our benefit. But the God of birds and trees would have to be also the God of birth defects and cancer.

Religious people have grappled for millennia with the theodicy, the problem posed by the existence of suffering in a world that is supposed to be ruled by a good God. They have found ingenious solutions in terms of various supposed divine plans. I will not try to argue with these solutions, much less to add one more of my own. Remembrance of the Holocaust leaves me unsympathetic to attempts to justify the ways of God to man.

If there is a God that has special plans for humans, then He has taken very great pains to hide His concern for us. To me it would seem impolite if not impious to bother such a God with our prayers.

Not all scientists would agree with my bleak view of the final laws. I do not know of anyone who maintains explicitly that there is scientific evidence for a divine being, but several scientists do argue for a special status in nature for intelligent life. Of course, everyone knows that as a practical matter biology and psychology have to be studied in their own terms, not in terms of elementary particle physics, but that is not a sign of any special status for life or intelligence; the same is true of chemistry and hydrodynamics. If, on the other hand, we found some special role for intelligent life in the final laws at the point of convergence of the arrows of explanation, we might well conclude that the creator who established these laws was in some way specially interested in us.

John Wheeler is impressed by the fact that according to the standard Copenhagen interpretation of quantum mechanics, a physical system cannot be said to have any definite values for quantities like position or energy or momentum until these quantities are measured by some observer's apparatus. For Wheeler, some sort of intelligent life is required in order to give meaning to quantum mechanics. Recently Wheeler has gone further and proposed that intelligent life not only must appear but must go on to pervade every part of the universe in order that every bit of information about the physical state of the universe should eventually be observed. Wheeler's conclusions seem to me to provide a good example of the dangers of taking too seriously the doctrine of positivism, that science should concern itself only with things that can be observed. Other physicists including myself prefer another, realist, way of looking at quantum mechanics, in terms of a wave function that can describe laboratories and observers as well as atoms and molecules, governed by laws that do not materially depend on whether there are any observers or not.

Some scientists make much of the fact that some of the fundamental constants have values that seem remarkably well suited to the appearance of intelligent life in the universe. It is not yet clear whether there is anything to this observation, but even if there is, it does not necessarily imply the operation of a divine purpose. In several modern cosmological theories, the so-called constants of nature (such as the masses of the elementary particles) actually vary from place to place or from time to time or even from one term in the wave function of the universe to another. If that were true, then as we have seen any scientists who study the laws of nature would have to be living in a part of the universe where the constants of nature take values favorable for the evolution of intelligent life.

For an analogy, suppose that there is a planet called Earth-prime, in every respect identical to our own, except that on this planet mankind developed the science of physics without knowing anything about astronomy. (E.g., one might imagine that Earthprime's surface is perpetually covered by clouds.) Just as on earth, students on Earthprime would find tables of fundamental constants at the back of their physics textbooks. These tables would list the speed of light, the mass of the electron, and so on, and also another "fundamental" constant having the value 1.99 calories of energy per minute per square centimeter, which gives the energy reaching Earthprime's surface from some unknown source outside. On earth this is called the solar constant because we know that this energy comes from the sun, but no one on Earthprime would have any way of knowing where this energy comes from or why this constant takes this particular value. Some physicist on Earthprime might note that the observed value of this constant is remarkably well suited to the appearance of life. If Earthprime received much more or much less than 2 calories per minute per square centimeter the water of the oceans would instead be vapor or ice, leaving Earthprime with no liquid water or reasonable substitute in which life could have evolved. The physicist might conclude that this constant of 1.99 calories per minute per square centimeter had been fine-tuned by God for man's benefit. More skeptical physicists on Earthprime might argue that such constants are eventually going to be explained by the final laws of physics, and that it is just a lucky accident that they have values favorable for life. In fact, both would be wrong. When the inhabitants of Earthprime finally develop a knowledge of astronomy, they learn that their planet receives 1.99 calories per minute per square centimeter because, like earth, it happens to be about 93 million miles away from a sun that produces 5,600 million million million million calories per minute, but they also see that there are other planets closer to their sun that are too hot for life and more planets farther from their sun that are too cold for life and doubtless countless other planets orbiting other stars of which only a small proportion are suitable for life. When they learn something about astronomy, the arguing physicists on Earthprime finally understand that the reason why they live on a world that receives roughly 2 calories per minute per square centimeter is just that there is no other kind of world where they *could* live. We in our part of the universe may be like the inhabitants of Earthprime before they learn about astronomy, but with other parts of the universe instead of other planets hidden from our view.

I would go further. As we have discovered more and more fundamental physical principles they seem to have less and less to do with us. To take one example, in the early 1920s it was thought that the only elementary particles were the electron and the proton, then considered to be the ingredients from which we and our world are made. When new particles like the neutron were discovered it was taken for granted at first that they had to be made up of electrons and protons. Matters are very different today. We are not so sure anymore what we mean by a particle being elementary, but we have learned the important lesson that the fact that particles are present in ordinary matter has nothing to do with how fundamental they are. Almost all the particles whose fields appear in the modern standard model of particles and interactions decay so rapidly that they are absent in ordinary matter and play no role at all in human life. Electrons are an essential part of our everyday world; the particles called muons and tauons hardly matter at all to our lives; yet, in the way that they appear in our theories, electrons do not seem in any way more fundamental than muons and tauons. More generally, no one has ever discovered any correlation between the importance of *anything* to us and its importance in the laws of nature.

Of course it is not from the discoveries of science that most people would have expected to learn about God anyway. John Polkinghorne has argued eloquently for a theology "placed within an area of human discourse where science also finds a home" that would be based on religious experience such as revelation, in much the way that science is based on experiment and observation. Those who think that they have had religious experiences of their own have to judge for themselves the quality of that experience. But the great majority of the adherents to the world's religions are relying not on religious experience of their own but on revelations that were supposedly experienced by others. It might be thought that this is not so different from the theoretical physicist relying on the experiments of others, but there is a very important distinction. The insights of thousands of individual physicists have converged to a satisfying (though incomplete) common understanding of physical reality. In contrast, the statements about God or anything else that have been derived from religious revelation point in radically different directions. After thousands of years of theological analysis, we are no closer now to a common understanding of the lessons of religious revelation.

There is another distinction between religious experience and scientific experiment. The lessons of religious experience can be deeply satisfying, in contrast to the abstract and impersonal worldview gained from scientific investigation. Unlike science, religious experience can suggest a meaning for our lives, a part for us to play in a great cosmic drama of sin and redemption, and it holds out to us a promise of some continuation after death. For just these reasons, the lessons of religious experience seem to me indelibly marked with the stamp of wishful thinking.

In my 1977 book, *The First Three Minutes,* I was rash enough to remark that "the more the universe seems comprehensible, the more it seems pointless." I did not mean that science teaches us that the universe is pointless, but rather that the universe itself suggests no point. I hastened to add that there were ways that we ourselves could invent a point for our lives, including trying to understand the universe. But the damage was done: that phrase has dogged me ever since. Recently Alan Lightman and Roberta Brawer published interviews with twenty-seven cosmologists and physicists, most of whom had been asked at the end of their interview what they thought of that remark. With various qualifications, ten of the interviewees agreed with me and thirteen did not, but of those thirteen three disagreed because they did not see why anyone would *expect* the universe to have a point. The Harvard astronomer Margaret Geller asked, ". . . Why should it have a point? What point? It's just a physical system, what point is there? I've always been puzzled by that statement." The Princeton astrophysicist Jim

Peebles remarked, "I'm willing to believe that we are flotsam and jetsam." (Peebles also guessed that I had had a bad day.) Another Princeton astrophysicist, Edwin Turner, agreed with me but suspected that I had intended the remark to annoy the reader. My favorite response was that of my colleague at the University of Texas, the astronomer Gerard de Vaucouleurs. He said that he thought my remark was "nostalgic." Indeed it was—nostalgic for a world in which the heavens declared the glory of God.

About a century and a half ago Matthew Arnold found in the withdrawing ocean tide a metaphor for the retreat of religious faith, and heard in the water's sound "the note of sadness." It would be wonderful to find in the laws of nature a plan prepared by a concerned creator in which human beings played some special role. I find sadness in doubting that we will. There are some among my scientific colleagues who say that the contemplation of nature gives them all the spiritual satisfaction that others have traditionally found in a belief in an interested God. Some of them may even really feel that way. I do not. And it does not seem to me to be helpful to identify the laws of nature as Einstein did with some sort of remote and disinterested God. The more we refine our understanding of God to make the concept plausible, the more it seems pointless.

Among today's scientists I am probably somewhat atypical in caring about such things. On the rare occasions when conversations over lunch or tea touch on matters of religion, the strongest reaction expressed by most of my fellow physicists is a mild surprise and amusement that anyone still takes all that seriously. Many physicists maintain a nominal affiliation with the faith of their parents, as a form of ethnic identification and for use at weddings and funerals, but few of these physicists seem to pay any attention to their nominal religion's theology. I do know two general relativists who are devout Roman Catholics; several theoretical physicists who are observant Jews; an experimental physicist who is a born-again Christian; one theoretical physicist who is a dedicated Moslem; and one mathematical physicist who has taken holy orders in the Church of England. Doubtless there are other deeply religious physicists whom I don't know or who keep their opinions to themselves. But, as far as I can tell from my own observations, most physicists today are not sufficiently interested in religion even to qualify as practicing atheists.

Religious liberals are in one sense even farther in spirit from scientists than are fundamentalists and other religious conservatives. At least the conservatives like the scientists tell you that they believe in what they believe because it is true, rather than because it makes them good or happy. Many religious liberals today seem to think that different people can believe in different mutually exclusive things without any of them being wrong, as long as their beliefs "work for them." This one believes in reincarnation, that one in heaven and hell; a third believes in the extinction of the soul at death, but no one can be said to be wrong as long as everyone gets a satisfying spiritual rush from what they believe. To borrow a phrase from Susan Sontag, we are surrounded by "piety without content." It all reminds me of a story that is told about an experience of Bertrand Russell, when in 1918 he was committed to prison for his opposition to the war. Following prison routine, a jailer asked Russell his religion, and Russell said that he was an agnostic. The jailer looked puzzled for a moment, and then brightened, with the observation that "I guess it's all right. We all worship the same God, don't we?"

Wolfgang Pauli was once asked whether he thought that a particularly ill-conceived physics paper was wrong. He replied that such a description would be too kind—the paper was not even wrong. I happen to think that the religious conservatives are wrong in what they believe, but at least they have not forgotten what it means really to believe something. The religious liberals seem to me to be not even wrong.

One often hears that theology is not the important thing about religion—the important thing is how it helps us to live. Very strange, that the existence and nature of God and grace and sin and heaven and hell are not important! I would guess that people do not find the theology of their own supposed religion important because they cannot bring themselves to admit that they do not believe any of it. But throughout history and in many parts of the world today people have believed in one theology or another, and for them it has been very important.

One may be put off by the intellectual muzziness of religious liberalism, but it is conservative dogmatic religion that does the harm. Of course it has also made great moral and artistic contributions. This is not the place to argue how we should strike a balance between these contributions of religion on one hand and the long cruel story of crusade and jihad and inquisition and pogrom on the other. But I do want to make the point that in striking this balance, it is not safe to assume that religious persecution and holy wars are perversions of true religion. To assume that they are seems to me a symptom of a widespread attitude toward religion, consisting of deep respect combined with a profound lack of interest. Many of the great world religions teach that God demands a particular faith and form of worship. It should not be surprising that *some* of the people who take these teachings seriously should sincerely regard these divine commands as incomparably more important than any merely secular virtues like tolerance or compassion or reason.

Across Asia and Africa the dark forces of religious enthusiasm are gathering strength, and reason and tolerance are not safe even in the secular states of the West. The historian Hugh Trevor-Roper has said that it was the spread of the spirit of science in the seventeenth and eighteenth centuries that finally ended the burning of witches in Europe. We may need to rely again on the influence of science to preserve a sane world. It is not the certainty of scientific knowledge that fits it for this role, but its *uncertainty*. Seeing scientists change their minds again and again about matters that can be studied directly in laboratory experiments, how can one take seriously the claims of religious tradition or sacred writings to certain knowledge about matters beyond human experience?

Of course, science has made its own contribution to the world's sorrows, but generally by giving us the means of killing each other, not the motive. Where the authority of science has been invoked to justify horrors, it really has been in terms of perversions of science, like Nazi racism and "eugenics." As Karl Popper has said, "It is only too obvious that it is irrationalism and not rationalism that has the responsibility for all national hostility and aggression, both before and after the Crusades, but I do not know of any war waged for a 'scientific' aim, and inspired by scientists."

Unfortunately I do not think that it is possible to make the case for scientific modes of reasoning by rational argument. David Hume saw long ago that in appealing to our past experience of successful science we are assuming the validity of the very mode of reasoning we are trying to justify. In the same way, all logical arguments can be defeated by the simple refusal to reason logically. So we cannot simply dismiss the question why, if we do not find the spiritual comfort we want in the laws of nature, we should *not* look for it elsewhere—in spiritual authority of one sort or another, or in an independent leap of faith?

The decision to believe or not is not entirely in our hands. I might be happier and have better manners if I thought I were descended from the emperors of China, but no effort of will on my part can make me believe it, any more than I can will my heart to stop beating. Yet it seems that many people are able to exert some control over what they believe and choose to believe in what they think makes them good or happy. The most interesting description I know of how this control can work appears in George Orwell's novel *1984*. The hero, Winston Smith, has written in his diary that "freedom is the freedom to say that two plus two is four." The inquisitor, O'Brien, takes this as a challenge, and sets out to force Smith to change his mind. Under torture Smith is perfectly willing to say that two plus two is five, but that is not what O'Brien is after. Finally, the pain becomes so unbearable that in order to escape it Smith manages to convince himself for an instant that two plus two *is* five. O'Brien is satisfied for the moment, and the torture is suspended. In much the same way, the pain of confronting the prospect of our own deaths and the deaths of those we love impels us to adopt beliefs that soften that pain. If we are able to manage to adjust our beliefs in this way, then why not do so?

I can see no scientific or logical reason not to seek consolation by adjustment of our beliefs—only a moral one, a point of honor. What do we think of someone who has managed to convince himself that he is bound to win a lottery because he desperately needs the money? Some might envy him his brief great expectations, but many others would think that he is failing in his proper role as an adult and rational human being, of looking at things as they are. In the same way that each of us has had to learn in growing up to resist the temptation of wishful thinking about ordinary things like lotteries, so our species has had to learn in growing up that we are not playing a starring role in any sort of grand cosmic drama.

Nevertheless, I do not for a minute think that science will ever provide the consolations that have been offered by religion in facing death. The finest statement of this existential challenge that I know is found in *The Ecclesiastical History of the English*, written by the Venerable Bede sometime around A.D. 700. Bede tells how King Edwin of Northumbria held a council in A.D. 627 to decide on the religion to be accepted in his kingdom, and gives the following speech to one of the king's chief men:

> Your majesty, when we compare the present life of man on earth with that time of which we have no knowledge, it seems to me like the swift flight of a single sparrow through the banqueting-hall where you are sitting at dinner on a winter's day with your thanes and counsellors. In the midst there is a comforting fire to warm the hall; outside, the storms of winter rain or snow are raging. This sparrow flies swiftly in through one door of the hall, and out through another. While he is inside, he is safe from the winter storms; but after a few moments of comfort, he vanishes from sight into the wintry world from which he came. Even so, man appears on earth for a little while; but of what went before this life or of what follows, we know nothing.

It is an almost irresistible temptation to believe with Bede and Edwin that there must be something for us outside the banqueting hall. The honor of resisting this temptation is only a thin substitute for the consolations of religion, but it is not entirely without satisfactions of its own.

CHAPTER XII

DOWN
IN ELLIS COUNTY

Mommas, don't let your babies grow up to be cowboys.
Don't let 'em pick guitars and drive them old trucks.
Make 'em be doctors and lawyers and such.

Ed and Patsy Bruce

Ellis County, Texas, is in the heart of what was once the greatest cotton-growing region in the world. It is not hard to find signs of the old cotton prosperity in Waxahachie, the county seat. The center of town boasts a grand 1895 pink granite county courthouse crowned with a high clock tower, and branching off from the central square are several streets of fine Victorian houses, looking like Cambridge's Brattle Street moved southwest. But the county is now much poorer. Although some cotton is still grown in the county along with wheat and corn, prices are not what they were. Dallas is forty minutes up Interstate 35 to the north, and a few prosperous Dallasites have moved into Waxahachie because they like the rural quiet, but the expanding aviation and computer industries of Dallas and Fort Worth have not come to Ellis County. By 1988 unemployment in Waxahachie stood at 7%. So it caused quite a stir around the county courthouse when on November 10, 1988, it was announced that Ellis County had been chosen as the site of the world's largest and most expensive scientific instrument, the Superconducting Super Collider.

Planning for the Super Collider had begun about six years earlier. At that time the Department of Energy had on its hands a troublesome project known as ISABELLE, already under construction at the Brookhaven National Laboratory on Long Island. ISABELLE had been intended as the successor to the existing Fermilab accelerator outside Chicago as America's leading facility for experimental research in elementary particle physics. After its start in 1978, ISABELLE had been set back two years by trouble in designing the superconducting magnets that would keep ISABELLE's beams of protons focused and on track. But there was another, deeper, problem with ISABELLE: although it would, when finished, be the most powerful accelerator in the world, it would probably not be powerful enough to answer the question that particle physicists most desperately needed to be answered: the question of how the symmetry that relates the weak and electromagnetic interactions is broken.

The description of the weak and electromagnetic forces in the standard model of elementary particles is based on an *exact* symmetry in the way that these forces enter into the equations of the theory. But as we have seen, this symmetry is not present in the solutions of the equations—the properties of the particles and forces themselves. Any version of the standard model that permits such a symmetry breakdown would have to contain features that have not yet been discovered experimentally: either new weakly interacting particles called Higgs particles, or new extra-strong forces. But we do not know which of these features is actually present in nature, and our progress in going beyond the standard model is blocked by this uncertainty.

The only sure way to settle this question is to do experiments in which a trillion volts is made available for the creation either of Higgs particles or of massive particles held together by extra-strong forces. For this purpose it turns out to be necessary to give a pair of colliding protons a total energy of about 40 trillion volts, because the proton energy is shared among the quarks and gluons of which the protons are composed, and only about one-fortieth of the energy would be available for the production of new particles in the collision of any one quark or gluon in one proton with a quark or gluon in the other. Furthermore, it is not enough to fire a beam of 40-trillion-volt protons into a stationary target because then almost all the energy of the incoming protons would be wasted in the recoil of the struck protons. In order to be confident of settling the question of the breakdown of the electroweak symmetry, one needs two beams of 20-trillion-volt protons that collide head-on, so that the momenta of the two protons cancel, and none of the energy has to be wasted in recoil. Fortunately one can be confident that an accelerator that produces intense colliding beams of 20-trillion-volt protons would in fact be able to settle the question of electroweak symmetry breaking—it would either find a Higgs particle, or it would find evidence of new strong forces.

In 1982 the idea began to circulate among experimental and theoretical physicists that the ISABELLE project ought to be scrapped and supplanted with construction of a much more powerful new accelerator that would be capable of settling the question of electroweak symmetry breaking. That summer an unofficial workshop of the American Physical Society carried out the first detailed study of an accelerator that would produce colliding beams of protons with energies of 20 trillion volts, about fifty times higher than planned for ISABELLE. In February of the following year a subpanel of the High Energy Physics Advisory Panel of the Department of Energy, under the leadership of Stanley Wojcicki of Stanford, began a series of meetings looking into the options for the next-generation accelerator. The subpanel met in Washington with the president's science adviser, Jay Keyworth, and received from him a strong hint that the administration would look favorably on a large new project.

The Wojcicki subpanel held its climactic meeting from June 29 to July 1, 1983, at Columbia University's Nevis Cyclotron Laboratory in Westchester County. The assembled physicists unanimously recommended the construction of an accelerator that could produce colliding beams of protons with energies of 10–20 trillion volts. In itself this vote would not have attracted so much attention; scientists in any field can generally be counted on to recommend new facilities for their research. Much more important was the ten-to-seven vote to recommend stopping work on ISABELLE. It was a dramatically difficult decision, one fought vigorously by Nick Samios, the director of Brookhaven. (Afterward Samios called this vote "one of the dumbest decisions ever made in high energy physics.") Not only did this decision dramatize the support of the subpanel for the big new accelerator—it made it politically very difficult for the Department of Energy to continue asking Congress for money for ISABELLE, and with ISABELLE stopped and nothing else beginning the Department of Energy would have no high-energy construction projects at all.

Ten days later the Wojcicki subpanel's recommendations were unanimously approved by its parent body, the Department of Energy's High Energy Physics Advisory Panel. Now for the first time the proposed new accelerator was given its present name: the Superconducting Super Collider, or SSC for short. On August 11 the Department of Energy authorized the High Energy Physics Advisory Panel to frame a plan for carrying out the research and development needed for the SSC project, and on November 16, 1983, Donald Hodel, the Secretary of Energy, announced his department's decision to stop work on ISABELLE and asked the appropriations committees of the House and Senate for authority to redirect funds from ISA-BELLE to the SSC.

The search for the mechanism for electroweak symmetry breaking was by no means the only motivation for the Super Collider. Usually new accelerators like those at CERN and Fermilab are built with the expectation that by moving to new levels of energy the accelerator will open up illuminating new phenomena. This expectation has almost always been fulfilled. For instance, the old Proton Synchotron was built at CERN with no definite idea of what it would find; certainly no one knew that experiments using neutrino beams from this accelerator would discover the neutral-current weak forces, a discovery that in 1973 verified our present unified theory of weak and electromagnetic forces. The large accelerators of today are descendants of Ernest Lawrence's Berkeley cyclotrons of the early 1930s, which were built to accelerate protons to a high enough energy to break through the electrical repulsion surrounding the atomic nucleus. Lawrence had no idea what would be found when the protons got inside the nucleus. Occasionally a particular discovery can be identified in advance; for instance, the Bevatron in Berkeley was built in the 1950s specifically to have enough energy (a mere 6 billion volts) to be able to create antiprotons, the antiparticles of the protons found in all ordinary atomic nuclei. The large electron-positron collider now operating at CERN was built primarily to have enough energy to produce very large numbers of Z particles and use them to subject the standard model to stringent experimental tests. But even where the design of a new accelerator is motivated by some specific problem, the most important discoveries it makes may be quite unexpected. This was certainly the case at the Bevatron; it did create antiprotons, but its most important achievement was in producing a great variety of unexpected new strongly interacting particles. In the same way, it was anticipated from the beginning that experiments at the Super Collider might make discoveries even more important than the mechanism for electroweak symmetry breaking.

Experiments at high-energy accelerators like the Super Collider may even solve the most important problem facing modern cosmology: the problem of the missing dark matter. We know that most of the mass of galaxies and an even larger fraction of the mass of clusters of galaxies is dark, not in the form of luminous stars like the sun. Even more dark matter is required in popular cosmological theories to account for the rate of expansion of the universe. This much dark matter could not be in the form of ordinary atoms; if it were, then the large numbers of neutrons and protons and electrons would affect calculations of the abundance of the light elements produced in the first few minutes of the expansion of the universe, so that the results of these calculations would no longer agree with observation.

So what is the dark matter? Physicists have been speculating for years about exotic particles of one sort or another that might make up the dark matter, but so far with no definite conclusions. If accelerator experiments revealed a new kind of long-lived particle, then by measuring its mass and interactions we would be able to calculate how many of these particles are left over from the big bang, and decide whether or not they make up all or part of the dark matter of the universe.

Recently these issues have been dramatized by observations made with the Cosmic Background Explorer (COBE) satellite. Sensitive microwave receivers on this satellite have discovered signs of tiny differences from one part of the sky to another in the temperature of radiation that is left over from a time when the universe was about three hundred thousand years old. It is believed that these nonuniformities in temperature arose from effects of the gravitational field that was produced by a slightly nonuniform distribution of matter at that time. This moment, three hundred thousand years after the big bang, was of critical importance in the history of the universe; the universe was then for the first time becoming transparent to radiation, and it is usually supposed that the nonuniformities in the distribution of matter were then just beginning to collapse under the influence of their own gravitation, ultimately to form the galaxies we see in the sky today. But the nonuniformities in the distribution of matter inferred from the COBE measurements are *not* young galaxies; COBE studied only irregularities of very large size, very much larger than the size that would have been occupied when the universe was three hundred thousand years old by the matter in a present-day galaxy. If we extrapolate what is seen by COBE down to the much smaller size of nascent galaxies, and in this way calculate the degree of nonuniformity of the matter at these relatively small scales, then we run into a problem: the galaxy-sized nonuniformities would have been too mild when the universe was three hundred thousand years old to have grown under the influence of their own gravitation into galaxies by the present. One way out of this problem is to suppose that nonuniformities of galactic size had already begun their gravitational condensation during the first three hundred thousand years, so that the extrapolation from what is seen by COBE down to the much smaller size of galaxies is not valid. But this is not possible if the matter of the universe is mostly composed of ordinary electrons, protons, and neutrons, because inhomogeneities in such ordinary matter could not have experienced any significant growth until the universe became transparent to radiation; at earlier times any clumps would have been blasted apart by the pressure of their own radiation. On the other hand, exotic dark matter that is composed of electrically neutral particles would have become transparent to radiation much earlier and would therefore have started its gravitational condensation much closer to the beginning, producing inhomogeneities at galactic scales much stronger than is inferred by extrapolation of the COBE results and strong enough, perhaps, to have grown into galaxies by the present. The discovery of a dark-matter particle produced at the Super Collider would validate this conjecture about the origin of galaxies, and would thus illuminate the early history of the universe.

There are many other new things that might be discovered at accelerators like the Super Collider: particles within the quarks that are within the protons; any of the various super-partners of known particles called for by supersymmetry theories; new kinds of force related to new internal symmetries; and so on. We do not know which if any of these things exist or whether, if they exist, they can be discovered at the Super Collider. It was therefore reassuring that we knew in advance of at least one discovery of high importance that the Super Collider could count on making, the mechanism for electroweak symmetry breaking.

After the Department of Energy's decision to build the SSC, there were years of planning and design ahead before construction could begin on the SSC. Long experience had shown that this sort of work, though sponsored by the federal government, is best carried on by private agencies, so the Department of Energy delegated the management of the research and development phase of the project to the Universities Research Association, a nonprofit consortium of sixty-nine universities that had been managing Fermilab. The association in turn recruited university and industrial scientists to serve as a board of overseers for the SSC, and we handed over the detailed job of designing the accelerator to a central design group at Berkeley headed by Maury Tigner of Cornell. By April 1986 the central design group had completed their design: a 10-foot-wide underground tunnel forming an 83-kilometer-long oval ring (comparable to the Washington Beltway), containing two slender beams of 20-trillion-volt protons traveling in opposite directions. The protons would be kept on their tracks by 3,840 bending magnets (each 17 meters long) and focused by 888 other magnets, the magnets containing altogether a total of 41,500 tons of iron, 19,400 kilometers of superconducting cable, and kept cool by 2 million liters of liquid helium.

On January 30, 1987, the project was approved by the White House. In April the Department of Energy began its site selection process by inviting proposals from interested states. By the deadline of September 2, 1987, they had received forty-three proposals (weighing altogether about 3 tons) from states that wanted to attract the SSC. A committee appointed by the National Academies of Science and Engineering narrowed the choice down to seven "best-qualified" sites, and then on November 10, 1988, the Secretary of Energy announced his department's decision: the SSC would go to Ellis County, Texas.

Part of the reason for this choice lies deep under the Texas countryside. Running north from Austin to Dallas is an eighty-million-year-old geological formation known as the Austin Chalk, laid down as sediment in a sea that covered much of Texas in the Cretaceous Period. Chalk is impervious to water, soft enough for easy digging, and yet strong enough so that it is unnecessary to reinforce the tunnel walls. One could hardly hope for better material through which to dig the Super Collider tunnel.

Meanwhile the struggle to fund the SSC was just beginning. A crucial moment for a project of this sort is the first appropriation for construction. Until that moment, the project is only a matter of research and development, which can be stopped as easily as it is started. Once construction begins, it becomes politically awkward to stop because stopping would amount to a tacit admission that the previously spent construction funds were wasted. In February 1988 President Reagan asked Congress for 363 million dollars in construction funds, but Congress appropriated only 100 million dollars and specifically tagged these funds for research and development, not construction.

The SSC project continued as if its future were assured. In January 1989 an industrial management team was selected, and Roy Schwitters of Harvard was chosen as director of the SSC Laboratory. Schwitters is a bearded but relatively young experimental physicist, then forty-four, who had proved his managerial abilities as leader of the major experimental collaboration at the leading U.S. high-energy facility, the Tevatron Collider at Fermilab. On September 7, 1989, we had some good news: a House-Senate conference committee agreed to appropriate 225 million dollars for the SSC in the 1990 fiscal year, with 135 million dollars for construction. The SSC project had finally gone beyond accelerator research and development.

The struggle was not over. Every year the SSC comes up again before Congress for funding, and every year the same arguments are made for and against it. Only a very naive physicist would be surprised at how little this debate has had to do with electroweak symmetry breaking or the final laws of nature. But only a very cynical physicist would fail to be a little saddened by the fact.

The single most powerful factor motivating politicians to support or oppose the SSC has been the immediate economic interests of their constituents. The project's congressional nemesis, Representative Don Ritter, has compared the SSC to "pork-barrel" projects that are pursued only for the political advantage of influential congressmen, calling it a "quark-barrel" project. Before the SSC site was chosen there was broad support for the project from those who hoped it would be located in their own states. When I testified in favor of the SSC before a Senate committee in 1987, one of the senators remarked to me that there were then almost one hundred senators in favor of the SSC but that after the site was announced there would be just two. Support certainly has shrunk, but the senator's estimate turned out to be overly pessimistic. Perhaps this is because companies throughout the country are receiving contracts for the components of the SSC, but I think that it also reflects some understanding of the intrinsic importance of the project.

Many of the opponents of the SSC point to the urgent need to reduce the federal deficit. This has been the recurrent theme of Senator Dale Bumpers of Arkansas, the leading opponent of the SSC in the Senate. I can understand this concern, but I do not understand why research at the frontier of science is the place to start reducing the deficit. One can think of many other projects, from the Space Station to the Sea Wolf submarine, whose cost is much greater than the SSC, and whose intrinsic value is far less. Is it to protect jobs that we must continue these other projects? But money spent on the SSC produces about as many jobs as the same money spent on anything else. Perhaps it is not too cynical to suggest that projects like the Space Station and the Sea Wolf submarine are too well protected politically by a network of aerospace and defense companies to be scrapped, leaving the SSC as a conveniently vulnerable target for a purely symbolic act of deficit reduction.

One of the persistent themes in the debate over the SSC was the argument over so-called big science versus small science. The SSC attracted opposition from some scientists who prefer an older and more modest style of science: experiments by a professor and a graduate student in the basement of a university building. Most of those who work in today's giant accelerator laboratories would also prefer physics in that style, but as a result of our past successes we now face problems that simply cannot be addressed with Rutherford's string and sealing wax. I imagine that many aviators are nostalgic for the days of open cockpits, but that is no way to cross the Atlantic.

Opposition to "big science" projects like the SSC also comes from scientists who would rather see the money spent on other research (such as their own). But I think that they are deluding themselves. When Congress has cut money from the administration budget request for the SSC, the freed funds have been allocated to water projects rather than to science. Many of these water projects are pure "pork," and they cost amounts that dwarf the funds being spent on the SSC.

The SSC also attracted opposition from those who suspect that President Reagan's decision to build the SSC was of a piece with his support for the "Star Wars" antimissile system and the space station: a mindless sort of enthusiasm for any large new technological project. On the other hand, much of the opposition to the SSC seemed to me to stem from an equally mindless distaste for any large new technological project. Journalists regularly lumped the SSC together with the space station as horrible examples of big science, despite the fact that the space station is not a scientific project at all. Arguing about big science versus small science is a good way to avoid thinking about the value of individual projects.

Some politically important support for the SSC comes from those who see it as a kind of industrial hothouse, forcing advances in various critical technologies: cryogenics, magnet design, on-line computing, and so on. The SSC would also represent an intellectual resource in helping our country to maintain a cadre of exceptionally gifted scientists. Without the SSC we shall lose a generation of high-energy physicists who will have to do their research in Europe or Japan. Even those who do not care about the discoveries made by these physicists may reflect that the community of high-energy physics represents a reservoir of scientific talent that has served our country well, from the Manhattan Project of the past to today's work on parallel programming for supercomputers.

These are good and important reasons for Congress to support the SSC, but they do not touch the heart of the physicist. The urgency of our desire to see the SSC completed comes from a sense that without it we may not be able to continue with the great intellectual adventure of discovering the final laws of nature.

I went down to Ellis County to look at the SSC site in the late autumn of 1991. Like much of this part of Texas the land is gently rolling, well watered by countless little creeks marked with stands of cottonwood trees. The ground was unlovely at that time of year; most of the crops had been harvested, and the fields planted to winter wheat were still only so much mud. Only here and there where harvesting had been delayed by recent rains a few fields were white with cotton. The sky was patrolled by hawks, hoping to catch a late-gleaning mouse. This is not cowboy country. I saw a huddled group of Black Angus cows and one white horse alone in a field, but the cattle that supply the stockyards of Fort Worth mostly come from ranches far to the north and west of Ellis County. On the way to the future SSC campus the good state farm-to-market roads dwindle down to unpaved county roads, no different from the dirt roads that served cotton farmers in this county a century ago.

I knew I had reached the land that Texas had bought for the SSC campus when I passed boarded-up farm houses, waiting to be moved or demolished. A mile or so to the north I could see an enormous new structure, the Magnet Development Building.

Then, beyond a copse of live-oak trees, I saw a tall drilling rig, brought up from the oil fields of the Gulf Coast in order to drill a 16-foot-wide test hole for the SSC, down 265 feet to the bottom of the Austin Chalk. I picked up a piece of chalk that had been dug up by the drill, and thought of Thomas Huxley.

Despite all the building and drilling, I knew that funding for the project might yet be stopped. I could imagine that the test holes might be filled in and the Magnet Building left empty, with only a few farmers' fading memories to testify that a great scientific laboratory had ever been planned for Ellis County. Perhaps I was under the spell of Huxley's Victorian optimism, but I could not believe that this would happen, or that in our time the search for the final laws of nature would be abandoned.

No one can say whether any one accelerator will let us make the last step to a final theory. I do know that these machines are necessary successors to a historical progression of great scientific instruments, extending back before the Brookhaven and CERN and DESY and Fermilab and KEK and SLAC accelerators of today to Lawrence's cyclotron and Thomson's cathode-ray tube and farther back to Frauenhofer's spectroscope and Galileo's telescope. Whether or not the final laws of nature are discovered in our lifetime, it is a great thing for us to carry on the tradition of holding nature up to examination, of asking again and again why it is the way it is.

THE SUPER COLLIDER,
ONE YEAR LATER

Just as this book was going to press in late October 1993, the House of Representatives voted to terminate the Superconducting Super Collider program. Although in the past the program has been saved after such votes, as of this writing the cancellation seems final. Political scientists and historians of science will doubtless find much employment in the years to come analyzing this decision, but it may not be too early to offer some comments on how it happened, and why.

On June 24, 1993, the House of Representatives voted to delete funding for the Super Collider from the energy and water appropriation bill, as they had done in 1992. This did not reduce the energy and water appropriation, or provide increased support for other areas of science; the funds for the Super Collider simply became available for other energy and water projects. Now only a favorable vote in the Senate could save the Laboratory.

Once again, physicists from all parts of the U.S. left their desks and laboratories to lobby in Washington throughout the summer in favor of the Super Collider. The theatrical high point of the struggle over the Super Collider's survival probably came in a Senate debate on September 29 and 30, 1993. Watching the debate, I had the surreal experience of hearing senators on the floor of the Senate arguing about the existence of Higgs bosons, and quoting this book as an authority. Finally, on September

30, the Senate voted 57–42 to fund the Super Collider at the full amount ($640 million) requested by the administration. This decision was then upheld by the House-Senate conference committee, but, on October 19, the House of Representatives voted by almost two to one to reject the conference committee's report, and to send the energy and water appropriations bill back to the committee with instructions to delete funding for the Super Collider. That committee has now met and agreed to terminate the project.

Why did this happen? Certainly the Super Collider program did not encounter any technical obstacles. In the year since this book was written, fifteen miles of the main tunnel have been dug through the Austin Chalk beneath the surface of Ellis County, Texas. The housing is completed and technical equipment partly installed for the linear accelerator, the first in a series of accelerators that was designed to start protons on their way through the Super Collider. Work is finished on the 570-meter tunnel of the Low Energy Booster, which would have accelerated protons emerging from the linear accelerator to twelve billion electron volts before passing them on to the Medium Energy Booster. (This is low energy by today's standards, but when I started physics research, twelve billion volts would have been beyond the capacity of any laboratory in the world.) Factories have been established in Louisiana, Texas, and Virginia to mass-produce the magnets that were to guide and focus protons on their way through the three boosters and the fifty-four-mile main ring. The Magnet Development Laboratory that I visited in 1991 has been joined on the site by other buildings—a Magnet Test Laboratory, an Accelerator Systems Test Building, and a building to house the massive refrigerators and compressors for the liquid helium needed to cool the superconducting magnets of the main ring. One experimental program—a collaboration of over a thousand

Ph.D. physicists from twenty-four different countries—was provisionally approved and another was close to approval.

Nor were there any discoveries in elementary particle physics that would have weakened the rationale for the Super Collider. In our efforts to go beyond the standard model, we are still stuck. Without the Super Collider, our best hope is that physicists in Europe will go ahead with their plans for a similar accelerator.

The troubles of the Super Collider project have been partly a side effect of unrelated political changes. President Clinton has continued the administration's support of the Super Collider, but he has less at stake in it politically than did President Bush of Texas, or President Reagan, during whose administration the project started. Perhaps most importantly, many members of Congress (especially new ones) now feel it necessary to show their fiscal prudence by voting against *something*. The Super Collider represents forty-three-thousandths of a percent of the federal budget, but it has become a convenient political symbol.

The most persistent note in the Super Collider debate was an expressed concern with priorities. This is a serious issue; seeing some of our fellow citizens ill-fed and ill-housed, it is never easy to spend money on other matters. But some members of Congress recognize that in the long run what our society gains from the support of basic science outweighs any immediate good that could be done with these funds. On the other side, many members of Congress who vigorously questioned the priority of spending for the Super Collider regularly vote for other projects that are far less worthwhile. Other larger projects, like the space station, have survived this year, less because of their intrinsic value than because so many constituents of members of Congress have an economic stake in these programs. Perhaps if the Super Collider had cost twice as much and created twice as many jobs, it would have fared better.

Opponents of the Super Collider also made much of charges of mismanagement and runaway costs. In fact, there was no mismangagement at the Super Collider and almost all cost increases have been due to delays in government funding. I said as much when I testified before the Senate Energy and Natural Resources Committee in August 1993. The best answer to these charges was the statement in August by Secretary of Energy O'Leary that after spending twenty percent of the total project cost, the Super Collider is now twenty percent complete.

Some members of Congress have argued that, although the Super Collider is scientifically valuable, we just can't afford it right now. But whenever we start a project of this size, there is bound to be some time during the years it takes to complete the project when the economy will be bad. What should we do— keep starting large projects, only to terminate them whenever there is a downturn in the economy? Now that we are writing off the two billion dollars and ten thousand man-years already invested in the Super Collider, what scientists or foreign governments would want to participate in any such project in the future, when it might be cancelled whenever the economy is bad again? Certainly any program should be reconsidered if changes in science or technology warrant it. Indeed, it was the high energy physicists who took the lead in cancelling ISABELLE, the last big accelerator project, when changes in physics goals made that appropriate. But there has been no change in the reasons for building the Super Collider. With the cancellation of the Super Collider program now, after all the work done on it, because the budget is tight this year, the United States seems permanently to be saying goodbye to any hope of ever having a responsible program of research in elementary particle physics.

Thinking back over the summer's struggle, I take some comfort from the observation that there are members of Congress who, apart from any political or economic motives they may have for supporting the Super Collider, are really interested in the science that it would do. One of them is Senator Bennett Johnston of Louisiana, who organized the pro–Super Collider side of the debate in the Senate. His home state had an important economic interest in building magnets for the Super Collider, but, beyond that, he is an enthusiastic science buff, as demonstrated in his eloquent speech on the Senate floor. The same intellectual excitement with science can be found in statements made by other members of Congress, such as Senators Moynihan of New York and Kerrey of Nebraska and Congressmen Nadler of Manhattan and Gephardt of Missouri and by the President's science advisor, Jack Gibbons. In May, I was one of the group of physicists who met with freshmen members of Congress. After others had explained the valuable technological experience to be gained by building the Super Collider, I remarked that though I did not know much about politics, I thought that one should not forget that there are many voters who are sincerely interested in the fundamental problems of science, apart from any applications to technology. A congressman from California then commented that he agreed with me about just one thing, that I didn't know much about politics. A little later, a congressman from Maryland came into the room, and after listening to the discussion of technological spin-offs for a while, remarked that one should not forget that many voters are interested, as well, in the fundamental problems of science. I went away happy.

The Super Collider debate also prompts reflections that are less cheerful. For centuries, the relations between science and society have been governed by a tacit bargain. Scientists generally want to make discoveries that are universal or beautiful or fundamental, whether or not they can foresee any specific benefit to society. Some people who are not themselves scientists find this sort of pure science exciting, but society, like the congressman from California, has generally been willing to support work in pure science mostly because it expects that it will yield applications. This expectation has generally proved correct. It is not that *any* work in science is liable occasionally to stumble onto something useful. Rather, it is just that when we push back the frontiers of knowledge we expect to find things that are really new, and that may be useful, in the way that radio waves and electrons and radioactivity have turned out to be useful. And the effort to make these discoveries also forces us into a sort of technological and intellectual virtuosity that leads on to other applications.

But now this bargain seems to be unraveling. It is not only some members of Congress who have lost confidence in pure science; the struggle for funds has led some of the scientists themselves, who work in more applied fields, to turn against support for those of us who search for the laws of nature. And the trouble that the Super Collider has faced in Congress is just one symptom of this disenchantment with pure science. Another is a recent attempt in the Senate to require that the National Science Foundation devote sixty percent of its spending toward societal needs. I don't say that the money would not be well spent, but it is appalling that some senators would choose research in pure science as the place from which to take these funds. The debate over the Super Collider has raised issues whose importance goes beyond the Super Collider itself, and that will be with us for decades to come.

Austin, Texas
October 1993

NOTES

CHAPTER I. PROLOGUE

8   Aristotle describes the motion of a projectile as being partly natural and partly unnatural: I had always thought that Aristotle taught that a projectile would travel in a straight line until its initial impulse was exhausted and then drop straight down, but I was unable to find this statement anywhere in his works. An expert on Aristotle, Robert Hankinson of the University of Texas, assures me that in fact Aristotle never said anything so contrary to observation and that this is a medieval misstatement of Aristotle's views.

10   Indeed, the word "law" was rarely used in antiquity: E. Zilsel, "The Genesis of the Concept of Physical Law," *Philosophical Review* 51 (1942): 245.

11   The classicist Peter Green blames the limitations of Greek science: Peter S. Green, *Alexander to Actium: The Historical Evolution of the Hellenistic Age* (Berkeley and Los Angeles: University of California Press, 1990), pp. 456, 475–78.

12   Newton described in the *Opticks* how he thought his program might be carried out: I am grateful to Bengt Nagel for suggesting the use of this quotation.

13   Robert Andrews Millikan, another American experimentalist: *The Autobiography of Robert A. Millikan* (New York: Prentice-Hall, 1950), p. 23. Also see a note by K. K. Darrow, *Isis* 41 (1950): 201.

13   A friend has told me that when he was a student at Cambridge: The physicist Abdus Salam.

13   there is plenty of other evidence for a ... sense of scientific complacency: The evidence for a sense of complacency in late-nineteenth-century science has been collected by the Berkeley historian Lawrence Badash, in "The Completeness of Nineteenth-Century Science," *Isis* 63 (1972): 48–58.

15   **"The day appears not far distant":** A. A. Michelson, *Light Waves and Their Uses* (Chicago: University of Chicago Press, 1903), p. 163.

16   **"the underlying physical laws":** P. A. M. Dirac, "Quantum Mechanics of Many Electron Systems," *Proceedings of the Royal Society* A123 (1929): 713.

17   **As his biographer Abraham Pais puts it:** Quoted by S. Boxer in the *New York Times Book Review,* January 26, 1992, p. 3.

CHAPTER II.   ON A PIECE OF CHALK

20   **The title of his lecture was:** Thomas Henry Huxley, *On a Piece of Chalk,* ed. Loren Eisley (New York: Scribner, 1967).

23   **so the absorption of these photons leaves the remaining reflected light greenish blue:** The precise color varies from one copper compound to another because the energies of atomic states are affected by the surrounding atoms.

25   **a list of open questions:** D. J. Gross, "The Status and Future Prospects of String Theory," *Nuclear Physics B* (Proceedings Supplement) 15 (1990): 43.

26   **ten examples of questions:** E. Nagel, *The Structure of Science: Problems in the Logic of Scientific Explanation* (New York: Harcourt, Brace, 1961).

27   **Newton derived his famous laws of motion in part from the earlier laws of Kepler:** According to Kepler's laws, the planets' orbits are ellipses with the sun at one focus; each planet's speed changes as it revolves around the sun in such a way that the line between the planet and the sun sweeps out equal areas in equal times; and the squares of the periods are proportional to the cubes of the largest diameters of the elliptical orbits. Newton's laws state that each particle in the universe attracts every other particle with a force proportional to the product of their masses and inversely proportional to the square of the distance, and dictate how any body moves under the influence of any given force.

28   "when state-of-the-art theoretical methods are intelligently applied": H. F. Shaefer III, "Methylene: A Paradigm for Computational Quantum Chemistry," *Science* 231 (1986): 1100.

28   we are not certain that we will ever know how to do these calculations: A number of theorists are pursuing the possibility of doing calculations involving strong nuclear forces by representing space-time as a lattice of distinct points, and using computers operating in parallel to keep track of the value of the fields at each point. It is hoped but not certain that by such methods the properties of nuclei could be deduced from the principles of quantum chromodynamics. So far it has not been possible even to calculate the masses of the proton and neutron, of which nuclei are composed.

28   "at the basis of the whole modern view of the world lies the illusion": L. Wittgenstein, *Tractatus Logico-Philosophicus,* trans. D. F. Pears and B. F. McGuiness (London: Routledge, 1922), p. 181. In much the same vein, a philosophically oriented friend, Professor Philip Bobbitt of the University of Texas School of Law, has commented to me, "When I say to a child who asks why an apple falls to earth, 'It's because of gravity, dear,' I am not explaining anything. The mathematical descriptions of the physical world that physics provides are not explanations. . . ." I agree with this if all that is meant by gravity is that there is a tendency of heavy objects to fall to the earth. On the other hand, if by gravity we understand the whole complex of phenomena described by Newton's or Einstein's theories, phenomena that include the motions of tides and planets and galaxies, then the answer that the apple falls because of gravity certainly feels to me like an explanation. At any rate this is how the word "explanation" is used by working scientists.

30   When the rules of quantum mechanics are applied to the atoms of which chalk is composed: The most stable elements are those with a number of electrons that can fit together neatly in complete shells; these are the noble gases, helium (two electrons), neon (ten electrons), argon (eighteen electrons), and so on. (These gases are called noble because as a result of the stability of their atoms they tend not to participate in chemical reactions.) Calcium has twenty electrons, so it has two outside the complete shells of argon, which it finds easy to lose. Oxygen has eight electrons, so it is two short of having enough for the complete shells of neon and readily picks up two electrons to fill the holes in its shells. Carbon has six electrons, so it can be regarded as either helium with four extra electrons, or neon with four missing, and so can either lose or gain four electrons. (This ambivalence allows carbon atoms to bind very strongly to each other, as in a diamond.)

30   **The number of protons must equal the number of electrons in order to keep the atom electrically neutral:** If the atom carries a positive or negative electric charge, then it tends to pick up or lose electrons until it becomes electrically neutral.

36   **the weird fossils of the Burgess Shale:** S. J. Gould, *Wonderful Life: The Burgess Shale and the Nature of History* (New York: Norton, 1989).

39   **The idea of emergence was well captured:** P. Anderson, *Science* 177 (1972): 393.

39   **a certain quantity called entropy:** To define entropy, imagine that some system's temperature is very slowly raised from absolute zero. The increase in entropy of the system as it receives each small new amount of heat energy is equal to that energy divided by the absolute temperature at which the heat is supplied.

39   **entropy, which always increases with time in any closed system:** It is important to note that the entropy may decrease in a system that can exchange energy with its environment. The emergence of life on earth represents a decrease of entropy, which is allowed by thermodynamics because the earth receives energy from the sun and loses energy to outer space.

40   **Ernest Nagel gave this as a paradigmatic example of the reduction of one theory to another:** E. Nagel, *The Structure of Science,* pp. 338–45.

41   **a battle was fought between the supporters of the new statistical mechanics and:** The story of this battle is told by the historian Stephen Brush in *The Kind of Motion We Call Heat* (Amsterdam: North-Holland, 1976), especially in section 1.9 of book 1.

41   **the explanation of why thermodynamics does apply to any particular system:** Thermodynamics applies to black holes, not because they contain a large number of atoms, but because they contain a large number of the fundamental mass units of the quantum theory of gravitation, equal to about one hundred thousandth of a gram and known as the Planck mass. It would not be possible to apply thermodynamics to a black hole that weighed less than a hundred thousandth of a gram.

43   "Most of the useful concepts of chemistry": R. Hoffman, "Under the Surface of the Chemical Article," *Angewandte Chemie* 27 (1988): 1597–1602.

43   some of the useful concepts of chemistry that were in danger of being lost: H. Primas, *Chemistry, Quantum Mechanics, and Reductionism,* 2nd ed. (Berlin: Springer-Verlag, 1983).

44   "There is no part of chemistry that does not depend": L. Pauling, "Quantum Theory and Chemistry," in *Max Plank Festschrift,* ed. B. Kockel, W. Mocke, and A. Papapetrou (Berlin: VEB Deutscher Verlag der Wissenschaft, 1959), pp. 385–88.

44   "What is surely impossible": A. B. Pippard, "The Invincible Ignorance of Science," Eddington Memorial Lecture delivered at Cambridge on January 28, 1988, *Contemporary Physics* 29 (1988): 393.

45   Where is one to draw the line? Sometimes it is argued that it is language that makes the difference between man and other animals and that humans only became conscious when they began to speak. Still, computers use language and do not seem to be conscious, but our old Siamese cat Tai Tai never spoke (and had a limited range of facial expressions) and yet in every other way showed the same signs of consciousness as humans.

45   "the ghost in the machine": G. Ryle, *The Concept of Mind* (London: Hutchinson, 1949).

46   "the words *realism* and *realist* might never again be used": G. Gissing, *The Place of Realism in Fiction,* reprinted in *Selections Autobiographical and Imaginative from the Works of George Gissing* (London: Jonathan Cape and Harrison Smith, 1929), p. 217.

48   Once in a television interview: B. Moyers, *A World of Ideas,* ed. B. S. Flowers (New York: Doubleday, 1989), pp. 249–62.

48   Similarly, when Philip Anderson recently wrote disparagingly: P. Anderson, "On the Nature of Physical Law," *Physics Today,* December 1990, p. 9.

48   "consciousness-related anomalous phenomena": R. G. Jahn and B. J. Dunne, *Foundations of Physics* 16(1986): 721. To be fair, I should add that Jahn sees his work as a reasonable extension of the Copenhagen interpretation of quantum mechanics, rather than as part of a paranormal agenda. The realist "many-histories" interpretation of quantum mechanics has the advantage of helping us to avoid this sort of confusion.

48 "although his . . . office is only a few hundred yards from my own": R. G. Jahn, letter to the editor, *Physics Today,* October 1991, p. 13.

49 much less on anything as small as a person: The general theory of relativity rests in large part on the principle that gravitational fields have *no* effects on a very small freely falling body, except to determine its falling motion. The earth is in free-fall in the solar system, so we on earth do not feel the gravitational field of the moon or sun or anything else, except for effects like tides that arise because the earth is not very small.

CHAPTER III.   TWO CHEERS FOR REDUCTIONISM

51 Canada's Science Council recently attacked: *Science,* August 9, 1991, p. 611.

52 It is nothing more or less than the perception that scientific principles are the way they are because of deeper scientific principles: Once in an article I called this view "objective reductionism"; see S. Weinberg, "Newtonianism, Reductionism, and the Art of Congressional Testimony," *Nature* 330 (1987): 433–37. I doubt that the phrase will catch on with philosophers of science, but it has been picked up by at least one biochemist, Joseph Robinson, in a response to an attack on reductionism by the philosopher H. Kincaid. See J. D. Robinson, "Aims and Achievements of the Reductionist Approach in Biochemistry/Molecular Biology/Cell Biology: A Response to Kincaid," *Philosophy of Science,* to be published.

52 Dostoevsky's underground man imagines a scientist: Fyodor Dostoevsky, *Notes from Underground,* trans. Mirra Ginsburg (New York: Bantam Books, 1974), p. 13.

53 It started when, in a 1985 article: E. Mayr, "How Biology Differs from the Physical Sciences," from *Evolution at a Crossroads,* ed. D. Depew and B. Weber (Cambridge, Mass.: MIT Press, 1985), p. 44.

53 he pounced on a line in a *Scientific American* article: S. Weinberg, "Unified Theories of Elementary Particle Interactions," *Scientific American* 231 (July 1974): 50.

53 I responded in an article: S. Weinberg, "Newtonianism."

53   There followed a frustrating correspondence: For some of this debate, see E. Mayr, "The Limits of Reductionism," and my reply, in *Nature* 331 (1987): 475.

54   "is perhaps the most divisive issue ever to confront the physics community": R. L. Park, *The Scientist,* June 15, 1987 (adapted from a talk at the symposium "Big Science/Little Science" at the American Physical Society annual meeting, May 20, 1987).

55   "in no sense more fundamental": P. W. Anderson, letter to the *New York Times,* June 8, 1986.

56   "[the] DNA revolution led a generation of biologists to believe": H. Rubin, "Molecular Biology Running into a Cul-de-sac?" letter to *Nature* 335 (1988): 121.

56   "to be sure the chemical nature of a number of black boxes": E. Mayr, *The Growth of Biological Thought: Diversity, Evolution, and Inheritance* (Cambridge, Mass.: Harvard University Press, 1982), p. 62.

59   Condensed matter physicists will doubtless eventually solve the problem of high-temperature superconductivity without any direct help from elementary particle physicists: I am using the word "direct" here because in fact there is a good deal of indirect help that different branches of physics offer each other. Part of it is an intellectual cross-fertilization; condensed matter physicists picked up one of their chief mathematical methods (the so-called renormalization group method) from particle physics, and particle physicists learned about the phenomenon called spontaneous symmetry breaking from condensed matter physics. In his testimony in favor of the Super Collider project in 1987 congressional committee hearings, Robert Schrieffer (who with John Bardeen and Leon Cooper was one of the founders of our modern theory of superconductivity) stressed that he had been led to some of his own work on superconductivity from his experience with the meson theories of elementary particle physics. (In a recent article, "John Bardeen and the Theory of Superconductivity," *Physics Today,* April 1992, p. 46, Schrieffer mentions that his 1957 guess at the quantum mechanical wave function for a superconductor was stimulated by thinking back to work on field theory by Sin-Itiro Tomonaga twenty years earlier.) Of course there are other ways that the different branches of physics help each other; for instance, the power demands of the Super Collider would make the project hopelessly too expensive if it were not possible to make magnets with superconducting cables;

and the synchrotron radiation emitted as a by-product in some high-energy particle accelerators has turned out to be of great value in medicine and in the study of materials.

60   **"I would therefore sharpen the criterion of scientific merit":** A. M. Weinberg, "Criteria for Scientific Choice," *Physics Today,* March 1964, pp. 42–48. Also see A. M. Weinberg, "Criteria for Scientific Choice," *Minerva* 1 (Winter 1963): 159–71; and "Criteria for Scientific Choice II: The Two Cultures," *Minerva* 3 (Autumn 1964): 3–14.

60   **an article of mine on these issues:** S. Weinberg, "Newtonianism."

60   **It was Gleick who introduced the physics of chaos to a general readership:** J. Gleick, *Chaos: Making a New Science* (New York: Viking, 1987).

60   **In a recent talk he argued:** Closing address by James Gleick at the 1990 Nobel Conference at Gustavus Adolphus College, October 1990.

CHAPTER IV.   QUANTUM MECHANICS AND ITS DISCONTENTS

71   **one number for each point of the space through which the wave passes:** Of course there is an infinite number of points in any volume of space, and it is not really possible to list the numbers representing any wave. But for purposes of visualization (and often in numerical calculations) it is possible to imagine space to consist of a very large but finite number of points, extending through a large but finite volume.

71   **The electron wave could also be described at any moment as a list of numbers:** These are actually complex numbers, in the sense that they generally involve the quantity symbolized by the letter i, equal to the square root of minus one, as well as ordinary numbers, both positive and negative. The part of any complex number proportional to i is called its imaginary part; the rest is called its real part. I will slur over this complication here because, important as it is, it does not really affect the points that I want to make about quantum mechanics.

72   **when such a wave packet strikes an atom, it breaks up:** Indeed, the wave packet of the electron begins to break up even before

the electron strikes the atom. Eventually this was understood to be due to the fact that according to the probabilistic interpretation of quantum mechanics the wave packet does not represent an electron with one definite velocity but with a distribution of different possible velocities.

73   **an electron wave that takes the form of a smooth, equally spaced alternation of crests and troughs extending over many wavelengths represents an electron that has a fairly definite momentum:** This description may give the misleading impression that in a state with a definite momentum, there is an alternation between points where the electron is not likely to be, where the corresponding values of the wave function are smallest, and points where the electron is most likely to be, where the values of the wave function are largest. This is not correct, because of the fact mentioned in an earlier endnote that the wave function is complex. There are really two parts of each value of the wave function, called its real and imaginary parts, which are out of phase with each other: when one is small, the other is large. The probability that an electron is in any particular small region is proportional to the sum of the squares of the two parts of the value of the wave function for that position, and this sum is strictly constant in a state of definite momentum.

74   **that he called *complementarity:*** N. Bohr, *Atti del Congresso Internazionale dei Fisici, Como, Settembre 1927*, reprinted in *Nature* 121 (1928): 78, 580.

75   **probabilities given by the squares of the values of the wave function:** Strictly speaking, it is the sum of the squares of the real and imaginary parts of the values of the wave function that give the probabilities of various configurations.

75   **We can think of this system as a mythical particle with only two ... positions:** Particles in the real world are of course not limited to only two positions, but there are physical systems that for practical purposes may be regarded as having just two configurations. The spin of an electron provides a real-world example of just such a two-state system. (The spin or angular momentum of any system is a measure of how fast it is spinning, how massive it is, and how far the mass extends from the axis of rotation. It is considered to have a direction, lying along the axis of rotation.) In classical mechanics the spin of a gyroscope or a planet could have any magnitude and direction. In contrast, in quantum mechanics if we measure the amount of an electron's spin

around any one direction, e.g., north (typically by measuring the energy of its interaction with a magnetic field in that direction), we can get only one of two results: the electron is spinning either clockwise or counterclockwise around that direction, but the magnitude of the spin is always the same: the magnitude of the electron's spin around any direction is equal to Planck's constant divided by 4π, or about a hundred millionth millionth millionth millionth millionth millionth millionth millionth millionth millionth the spin of the earth on its axis.

76   the probability that it will turn out to be *here* is given by the square of the *here* value . . . and the probability that it will turn out to be *there* is given by the square of the *there* value: The sum of these two probabilities must be one (i.e., 100%), so the sum of the squares of the *here* and *there* values must equal one. This suggests a very useful geometric picture. Draw a right triangle, with a horizontal side having a length equal to the *here* value of the wave function and a vertical side having a length equal to the *there* value. (Of course, by horizontal and vertical I just mean any two perpendicular directions. I could just as well have said uptown and crosstown.) You do not have to be a modern major general to know one cheerful fact about the square of the hypotenuse of this triangle: it equals the sum of the squares of the vertical and horizontal sides. But as we have seen, this sum has the value one, so the hypotenuse has length one. (By one I do not mean 1 meter or 1 foot, because probabilities are not measured in squared meters or squared feet; I mean the pure number one.) Conversely, if we are given an arrow of unit length with some definite direction in two dimensions (in other words, a two-dimensional unit vector), then its projection on the horizontal and vertical directions or any other pair of perpendicular directions gives a pair of numbers whose squares necessarily add up to one. Thus instead of specifying a *here* value and a *there* value, the state may just as well be represented by an arrow (the hypotenuse of our triangle) of length equal to one, whose projection on any direction is the value of the wave function for the configuration of the system that corresponds to that direction. This arrow is called the *state vector.* Dirac developed a somewhat abstract formulation of quantum mechanics in terms of state vectors, which has advantages over the formulation in terms of wave functions because we can speak of a state vector without reference to any particular configurations of the system.

76   the nature of the system in question: Of course, most dynamical systems are more complicated than our mythical particle. For instance, consider two such particles. There are here four possible configurations, in which particles one and two are respectively *here* and *here, here* and *there, there* and *here,* and *there* and *there.* Thus the wave function for the state of two such particles has four values, and there are sixteen constant numbers that are needed to describe how it evolves with time. Note that there is still just one wave function, describing the joint state of the two particles. This is generally the case; we do not have a separate wave function for each electron or other particle, but only one wave function for any system, however many particles it may contain.

76   **there is a pair of states . . . of definite momentum, that we may call *stop* and *go:*** In saying that these states have definite momentum, I am speaking loosely. With just two possible positions, the *go* state is as close as we can come to a smooth wave with a crest *here* and a trough *there,* corresponding to a particle with a non-zero momentum, while the *stop* state is like a flat wave, for which the wavelength is much larger than the distance from *here* to *there,* corresponding to a particle at rest. This is a primitive version of what mathematicians call Fourier analysis. (Strictly speaking we must take the *stop* and *go* values of the wave function as the sum or difference of the *here* and *there* values divided by the square root of two, in order to satisfy the condition mentioned in a previous endnote, that the sum of the squares of the two values must be equal to one.)

77   **writers like Fritjof Capra:** F. Capra, *The Tao of Physics* (Boston: Shambhala, 1991).

78   **too simple to allow chaotic solutions:** Physicists sometimes use the term "quantum chaos" to refer to the properties of quantum systems that *would* be chaotic in classical physics, but the quantum systems themselves are never chaotic.

81   **experimental physicists have demonstrated:** Notably Alain Aspect.

83   **each of these two histories will thenceforth unfold without interaction with the other:** The phenomenon by which the two histories of the world cease to interfere with each other is known as "decoherence." The study of how this comes about has attracted much attention lately, from theorists including Murray Gell-Mann and James Hartle, and independently Bryce De Witt.

84   **A long line of physicists have worked to purge the foundations of quantum mechanics of any statement about probabilities:**

Here is a partial list of references: J. B. Hartle, "Quantum Mechanics of Individual Systems," *American Journal of Physics* (1968): 704; B. S. De Witt and N. Graham, in *The Many-Worlds Interpretation of Quantum Mechanics* (Princeton: Princeton University Press, 1973), pp. 183–86; D. Deutsch, "Probability in Physics," Oxford University Mathematical Institute preprint, 1989; Y. Aharonov, paper in preparation.

88 **the nonlinearities of the generalized theory** *could* **be used to send signals instantaneously:** Polchinski subsequently found a slightly modified interpretation of this theory in which this sort of faster-than-light communication was forbidden, but in which the "different worlds" corresponding to different results of measurements can go on communicating with one another.

CHAPTER V. TALES OF THEORY AND EXPERIMENT

91 **the elliptical orbits . . . actually precess:** I.e., the orbits do not exactly close; a planet, in going from its point of closest approach to the sun, known as the perihelion, out to the point of farthest distance from the sun and then back to the point of closest approach goes a little bit more than 360 degrees around the sun. The resulting slow change of orientation of the orbit is thus usually called the precession of the perihelia.

95 **In a report to the 1921 Nobel committee:** Information quoted here about the Nobel Prize reports and nominations is taken from the superb scientific biography of Einstein by A. Pais, *Subtle Is the Lord: The Science and Life of Albert Einstein* (New York: Oxford University Press, 1982), chap. 30.

96 **Indeed, the astronomers of the 1919 eclipse expedition were accused of bias:** For a discussion and references, see D. G. Mayo, "Novel Evidence and Severe Tests," *Philosophy of Science* 58 (1991): 523.

96 **in the case of general relativity a** *retrodiction* **. . . provided a more reliable test of the theory than a true** *prediction:* I made this remark in my Bampton Lectures at Columbia University in 1984. Subsequently I was glad to see that the same conclusion was reached independently by a credentialed historian of science, Stephen Brush, in "Prediction and Theory Evaluation: The Case of Light Bending," *Science* 246 (1989): 1124.

97 **we have seen that the early experimental evidence for general relativity:** I should mention that Einstein had proposed a third test of general relativity based on a predicted gravitational redshift of light. Just as a projectile thrown up from the surface of the earth loses speed as it climbs up out of the earth's gravity, so also a light ray emitted from the surface of a star or planet loses energy as it climbs up to outer space. For light, this loss of energy is manifested as an increase of wavelength and hence (for visible light) a shift toward the red end of the spectrum. This fractional increase of wavelength is predicted by general relativity to be 2.12 parts per million for light from the surface of the sun. It was proposed to examine the spectrum of light from the sun, to see whether the spectral lines were shifted toward the red by this amount from their normal wavelengths. This effect was searched for by astronomers but not at first found, a fact that seems to have worried some physicists. The report of the 1917 Nobel committee noted that measurements of C. E. St. John at Mount Wilson had not found the redshift and concluded, "It appears that Einstein's Relativity Theory, whatever its merits in other respects may be, does not deserve a Nobel Prize." The 1919 Nobel committee report again mentioned the redshift as a reason for reserving judgment on general relativity. However, most physicists at the time (including Einstein himself) do not seem to have been greatly concerned with the redshift problem. Today we can see that the techniques used around 1920 could not have yielded an accurate measurement of the solar gravitational redshift. For instance, the predicted gravitational redshift of two parts per million could be masked by a shift produced by the convection of the light-emitting gases on the surface of the sun (the familiar Doppler effect) having nothing to do with general relativity. If these gases were rising toward the observer at a speed of 600 meters per second (not an impossible speed on the sun) the gravitational redshift would be entirely canceled. It is only in recent years that careful study of light from the edge of the sun's disk (where convection would be mostly at right angles to the line of sight) has revealed a gravitational redshift of about the expected magnitude. Indeed, the first precise measurements of the gravitational redshift used not light from the sun, but gamma rays (light of very short wavelength) that were allowed to rise or fall a mere 22.6 meters in the tower of the Jefferson Physical Laboratory at Harvard. A 1960 experiment by R. V. Pound and G. A. Rebka found a change in gamma-ray wavelength in agreement with general relativity to within a 10% experimental uncertainty, an accuracy improved a few years later to about 1%.

97   new techniques of radar and radio astronomy led to a significant improvement in the accuracy of these experimental tests: Especially in the work of Irwin Shapiro, then at MIT.

98   the motion of small particles in fluids: This is known as
Brownian motion. It is caused by the impact of molecules of the liquid
striking the particles. With the aid of Einstein's theory of Brownian
motion, observations of this motion could be used to calculate some
of the properties of molecules and also helped to convince chemists
and physicists of the reality of molecules.

99   Einstein tried at least one of them: For the experts, I refer here
to the massless scalar theory.

101   there is no *one* freely falling frame of reference: E.g., suppose
that we adopt a frame of reference that throughout all space is accelerating along the direction from Texas to the center of the earth at 32
feet per second per second. In this frame of reference we in Texas
would not feel a gravitational field because this is the frame of reference that is freely falling in Texas, but our friends in Australia would
feel twice the normal gravitational field, because in Australia this
frame of reference would be accelerating away from the center of the
earth, not toward it.

105   in Newton's theory one could have replaced the inverse-
square law with an inverse-cube law or an inverse 2.01-power law
without the slightest change in the conceptual framework of the
theory: This is true of Newton's formulation of his theory in terms of
a force acting at a distance, but not of the subsequent reformulation of
Newton's theory (by Laplace and others) as a field theory. But even in
the field-theoretic version of Newton's theory it would be easy to add
a new term to the field equations that would produce other changes in
the dependence of force on distance. Specifically, the inverse-square
law could be replaced with a formula that gives the gravitational force
an approximate inverse-square law behavior out to some distance, beyond which the force would fall off exponentially fast. This sort of
modification is not possible in general relativity.

107   the energy and momentum of the electric and magnetic fields
in a ray of light come in bundles: Born, Heisenberg, and Jordan actually considered only a simplified version of an electromagnetic field, in
which complications resulting from the polarization of light are ignored. These complications were considered a little later by Dirac, and

then a complete treatment of the quantum field theory of electromagnetism was given by Enrico Fermi.

108   may be calculated by adding up an infinite number of contributions: The allowed photon energies form a continuum, so this "sum" is actually an integral.

110   In the end, the solution to the problem of infinities that emerged in the late 1940s: The history of these developments is told by T. Y. Cao and S. S. Schweber, "The Conceptual Foundations and Philosophical Aspects of Renormalization Theory," to be published in *Synthèse* (1992).

111   Lamb had just succeeded in measuring: Strictly speaking, Lamb measured the difference in the energy shift of two states of the hydrogen atom that according to the old Dirac theory in the absence of photon emissions and reabsorptions should have exactly the same energy. Although Lamb could not measure the precise energies of these two atomic states, he could detect that their energies actually differed by a tiny amount, thus showing that something had shifted the energies of the two states by different amounts.

112   Is it possible that these two infinities cancel and leave a finite total energy?: This idea had been suggested some time earlier by Dirac, by Weisskopf, and by H. A. Kramers.

113   more accurate calculations of the Lamb shift: These calculations were done by Lamb himself with Norman Kroll, and by Weisskopf with J. B. French.

113   As Nietzsche says: From "Aus dem Nachlass der Achtzigerjahre," in a notebook from the 1880s published posthumously in F. Nietzsche, *Werke III,* ed. Schlecta, 6th ed. (Munich: Carl Hauser, 1969), p. 603. This remark is the theme of a novel, *Death of a Beekeeper* (New York: New Directions, 1981), by my Texas colleague Lars Gustafsson.

114   This calculation has been continually refined: These theoretical and experimental results are reviewed by T. Kinoshita, in *Quantum Electrodynamics,* ed. T. Kinoshita (Singapore: World Scientific, 1990).

115   I did not see what was so terrible about an infinity in the bare mass and charge: There is a more serious problem with quantum electrodynamics. In 1954 Murray Gell-Mann and Francis Low showed that the effective charge of the electron increases very slowly with the energy of the process in which it is measured, and they raised the possibility that (as guessed earlier by the Soviet physicist Lev Landau) the effective charge actually becomes infinite at some very high energy. More recent calculations have indicated that this disaster does occur in pure quantum electrodynamics, the theory of photons and electrons and nothing else. However the energy at which this infinity occurs is so high (much larger than that contained in the total mass of the observable universe) that long before such energies are reached it becomes impossible to ignore all the other sorts of particles in nature besides photons and electrons. Insofar as there is still any question of the mathematical consistency of quantum electrodynamics, it has merged with the question of the consistency of our quantum theories of all particles and forces.

117   the Fermi theory of the weak nuclear force was put in its final form: By Feynman and Gell-Mann and independently by Robert Marshak and George Sudarshan.

118   an analogy with quantum electrodynamics: I am referring here to the generalization of quantum electrodynamics by C. N. Yang and R. L. Mills.

121   In 1967 it received zero citations: This is not strictly true, because I mentioned this paper in a talk I gave at the Solvay Conference in Brussels in 1967. However, the ISI only counts published journal articles, and my remark was published in a conference proceeding.

121   the most frequently cited article: Eugene Garfield, "The Most-Cited Papers of All Time, SCI 1945–1988," in *Current Contents,* February 12, 1990, p. 3. To be more precise, it was the only article on elementary particle physics (or on any other sort of physics except for biophysics, chemical physics, and crystallography) in the one hundred articles in all the sciences that were most frequently cited during the period covered by the ISI survey, from 1945 to 1988. (Because of the war, there presumably were no frequently cited articles on elementary particle physics from 1938 to 1945.)

127   the physicists in Oxford and Seattle have repeated their experiments: I happened to be at Oxford a few years ago, and took the opportunity to ask Pat Sanders, who had led the Oxford experiment on bismuth, whether his group had ever figured out what had gone wrong with the earlier experiment. He told me that they had not and that unfortunately they never will because the Oxford experimentalists had cannibalized the apparatus and were using it as part of a new apparatus, which was now getting the right answer. So it goes.

128   **predicted a new kind of particle:** This was on the basis of a symmetry principle proposed by Roberto Peccei and Helen Quinn.

128   **The idea either is incorrect or needs modification:** Such modifications have been proposed by M. Dine, W. Fischler, and M. Srednicki, and by J. E. Kim.

129   **the discovery of a universal background of radio static:** By Arno Penzias and Robert Wilson. I have written about the discovery of this universal background in *The First Three Minutes: A Modern View of the Origin of the Universe* (New York: Basic Books, 1977).

131   **military historians often write as if generals lose battles because they do not follow some well-established rules of military science:** One example is Basil Liddell Hart, the advocate of the "indirect approach."

131   **This is called the art of war:** I must acknowledge that as the phrase "art of war" appears in the translations of the classic works of Sun Tzu, Jomini, and Clausewitz, the word "art" is used in opposition to "science" as "technique" is opposed to "knowledge," but not as "subjective" is opposed to "objective," or "inspiration" is opposed to "system." These authors' use of the term "art" served to emphasize that they were writing about the art of war because they wanted to be of use to people who actually would win wars, but they intended to go about this in a scientific and systematic way. The Confederate General James Longstreet used the term "art of war" in something like the sense that I use it here when he said that both McClellan and Lee were "masters of the science, but not of the art of war." (James Longstreet, *From Manassas to Appomattox* [Philadelphia: Lippincott, 1896], p. 288.) Later historians like Charles Oman and Cyril Falls who write of an "art of war" make clear that there is no system of war. The reader who has come this far will understand that there is not much of a system of science either.

CHAPTER VI.   BEAUTIFUL THEORIES

132   **the search for beauty in physics:** The astrophysicist Subrahmanyan Chandrasekhar has written movingly of the role of beauty in science, in *Truth and Beauty: Aesthetics and Motivations in Sci-*

ence (Chicago: University of Chicago Press, 1987), and *Bulletin of the American Academy of Arts and Sciences* 43, no. 3 (December 1989): 14.

135   in Einstein's theory there are fourteen: I am referring to the ten field equations plus the four equations of motion.

135   As Einstein said of general relativity: Quoted by G. Holton, "Constructing a Theory: Einstein's Model," *American Scholar* 48 (summer 1979): 323.

141   bundles called gravitons, that also behave like particles of zero mass: Gravitons have not been detected experimentally, but this is no surprise; calculations show that they interact so weakly that individual gravitons could not have been detected in any experiment yet performed. Nevertheless, there is no serious doubt of the existence of gravitons.

145   families of these other particle types: Strictly speaking, it is only the left-handed states of the electron and neutrino and the up-and-down quarks that form these families. (By left-handed, I mean that the particle spins in the direction that your fingers curl if the thumb of your left hand is laid along the axis of rotation of the particle pointing in the particle's direction of motion.) This distinction between the families formed by left- and right-handed states is the origin of the fact that the weak nuclear forces do not respect the symmetry between right and left. (The asymmetry between right and left in the weak forces was proposed in 1956 by the theorists T. D. Lee and C. N. Yang. It was verified by experiments on nuclear beta decay by C. S. Wu in collaboration with a group at the National Bureau of Standards in Washington, and in experiments of the decay of the pi meson by R. L. Garwin, L. Lederman, and M. Weinrich and by J. Friedman and V. Telegdi.) We still do not know why it is only the left-handed electrons and neutrinos and quarks that form these families; this is a challenge for theories that aim at going beyond our standard model of elementary particles.

146   reasons that are purely historical: In 1918 the mathematician Hermann Weyl proposed that the symmetry of general relativity under space-time-dependent changes of position or orientation should be supplemented by a symmetry under space-time-dependent changes in the way one measures (or "gauges") distances and times. This symmetry principle was soon abandoned by physicists (though versions of it crop up now and then in speculative theories), but it is mathematically very similar to an internal symmetry of the equations of electrodynamics, which therefore came to be called gauge invariance. Then, when a more complicated sort of local internal symmetry was introduced in 1954 by C. N. Yang and R. L. Mills (in an attempt to account for the strong nuclear force), it, too, was called a gauge symmetry.

146 **the internal property of quarks that is fancifully known as** *color:* Various versions of the attribute of quarks known as color were suggested by O. W. Greenberg; M. Y. Han and Y. Nambu; and W. A. Bardeen, H. Fritzsch, and M. Gell-Mann.

148 **the theory must be "renormalizable":** But see the remarks in chap. 8 qualifying this requirement.

152 **processes like nuclear beta decay that could not be understood along the lines of Dirac's theory:** In Dirac's theory electrons are eternal; a process like the production of an electron and a positron is interpreted as the lifting of a negative-energy electron to a state of positive energy, leaving a hole in the sea of negative-energy electrons that is observed as a positron, and the annihilation of an electron and positron is interpreted as the falling of an electron into such a hole. In nuclear beta decay electrons are created *without positrons* out of the energy and the electric charge in the electron field.

152 **there are other particles with other spins:** Dirac and I were at a conference in Florida in the early 1970s, and I took the occasion to ask him how he could explain the fact that there are particles (like the pi meson or the W particle) that have a spin different from the electron's and could not have stable states of negative energy, and yet have distinct antiparticles. Dirac said that he had never thought that these particles were important.

152 **A possible explanation was given by Niels Bohr:** This is a recollection of Heisenberg, quoted by Valentine Telegdi and Victor Weisskopf in a review of Heisenberg's collected works in *Physics Today,* July 1991, p. 58. The same notion of the limited variety of possible mathematical forms has been expressed by the mathematician Andrew Gleason.

153 **The English mathematician G. H. Hardy:** Throughout his life Hardy boasted that his research in pure mathematics could not possibly have any practical application. But when Kerson Huang and I were working at MIT on the behavior of matter at extremely high temperature, we found just the mathematical formulas we needed in Hardy's papers with Ramanujan on number theory.

153   **Carl Friedrich Gauss and others developed a non-Euclidean geometry:** The other principal architects of this curved space were Janos Bolyai and Nicolai Ivanovitch Lobachevski. The work of Gauss, Bolyai, and Lobachevski was important for the future of mathematics because they described this space as being not merely curved the way the surface of the earth is curved, by the way that the surface is embedded in a higher dimensional uncurved space, but in terms of its intrinsic curvature, without any reference to how it is embedded in higher dimensions.

153   **all Euclid's postulates except the fifth:** Euclid's fifth postulate in one version states that through any given point outside any given line, one and only one line can be drawn that is parallel to the given line. In the new non-Euclidean geometry of Gauss, Bolyai, and Lobachevski, many such parallel lines can be drawn.

155   **experiments in 1936 revealed that the nuclear force:** These experiments were carried out by Merle Tuve together with N. Heydenberg and L. R. Hafstad, using a million-volt Van de Graff accelerator to fire a beam of protons into a proton-rich target like paraffin.

155   **These symmetry transformations act ... in a way that is mathematically the same as the way that ordinary rotations in three dimensions act on the spins of particles:** For this reason, this symmetry is known as *isotopic spin symmetry*. (It was proposed in 1936 by G. Breit and E. Feenberg, and independently by B. Cassen and E. U. Condon, on the basis of the experiments of Tuve et al.) Isotopic spin symmetry is also mathematically similar to the internal symmetry that underlies the weak and electromagnetic forces in the electroweak theory, but physically quite different. One difference is that different particles are grouped into families: the proton and neutron for isotopic spin symmetry, and the left-handed electron and neutrino as well as left-handed up and down quarks for the electroweak symmetry. Also, the electroweak symmetry states the invariance of the laws of nature under transformations that can depend on position in space and time; the equations governing nuclear physics preserve their form only if we transform protons and neutrons into each other in the same way everywhere and at all times. Finally the isotopic spin symmetry is only approximate and is understood today as a somewhat accidental consequence of the small masses of quarks in our modern theory of strong nuclear forces; the electroweak symmetry is exact and taken as a fundamental principle in the electroweak theory.

156   **a mathematical structure known as a *group*:** If two transformations leave something unchanged then so does their "product," defined by performing one transformation and then the other. If a transformation leaves something unchanged, then so does its inverse, the transformation that undoes the first. Also, there is always one transformation that leaves anything unchanged, the transformation that does nothing at all, known as the unit transformation because it acts like multiplication by the number one. These three properties are what make any set of operations a group.

156   **a list of all the "simple" Lie groups:** Briefly, there are three infinite categories of simple Lie groups: the familiar rotation groups in two, three, or more dimensions, and two other categories of transformations somewhat like rotations, known as unitary and symplectic transformations. In addition there are just five "exceptional" Lie groups that do not belong to any of these categories.

157   **this predicted particle was subsequently discovered in 1964:** By a group headed by N. Samios.

157   **there are no general formulas for the solution of certain algebraic equations:** The group in question in Galois's work was the set of permutations of the solutions of the equation.

157   **A well-known essay by the physicist Eugene Wigner:** E. P. Wigner, "The Unreasonable Effectiveness of Mathematics," *Communications in Pure and Applied Mathematics* 13 (1960): 1–14.

160   **It was not until the development of a rigorous and abstract mathematical style:** J. L. Richards, "Rigor and Clarity: Foundations of Mathematics in France and England, 1800–1840," *Science in Context* 4 (1991): 297.

162   **Francis Crick describes in his autobiography:** F. Crick, *What Mad Pursuit: A Personal View of Scientific Discovery* (New York: Basic Books, 1988).

163   **some triplets produce nothing at all:** Strictly speaking, the otherwise meaningless triplets do carry the message "end chain."

164   **Kepler wrote that he had:** In a May 1605 letter from Kepler to Fabricius, quoted by E. Zilsel, "The Genesis of the Concept of Physical Law," *Philosophical Review* 51 (1942): 245.

CHAPTER VII. AGAINST PHILOSOPHY

167   **This is not to deny all value to philosophy:** Two philosopher friends have pointed out to me that this chapter's title, "Against Philosophy," is an exaggeration, because I am not arguing against philosophy in general but only against the bad effects on science of philosophical doctrines like positivism and relativism. They speculated that I intended the title as a response to Feyerabend's book, *Against Method*. Actually the title of this chapter was suggested to me by the titles of a pair of well-known law review articles, Owen Fiss's "Against Settlement" and Louise Weinberg's "Against Comity." Anyway, I did not think that "Against Positivism and Relativism" would be a very catchy title.

167   **"these almost arcane discussions":** G. Gale, "Science and the Philosophers," *Nature* 312 (1984): 491.

167   **"nothing seems to me less likely":** L. Wittgenstein, *Culture and Value* (Oxford: Blackwell, 1980).

168   **Some of it I found to be written in a jargon so impenetrable:** For examples, see some of the articles in *Reduction in Science: Structure, Examples, Philosophical Problems*, ed. W. Balzer, D. A. Pearce, and H.-J. Schmidt, (Dordrecht: Reidel, 1984).

168   **But only rarely did it seem to me to have anything to do with the work of science:** Many other working scientists have the same reaction to the writings of philosophers. E.g., in his reply to the philosopher H. Kincaid that I quoted in chap. 3, the biochemist J. D. Robinson remarked that "biologists undoubtedly commit heinous philosophical sins. And they ought to welcome enthusiastically the informed attention of philosophers. That attention, however, will be most useful when philosophers recognize what biologists intend and what biologists do."

168   **According to Feyerabend:** P. K. Feyerabend, "Explanation, Reduction, and Empiricism," *Minnesota Studies in the Philosophy of Science* 3 (1962): 46–48. The philosophers to which Feyerabend refers are the positivists of the Vienna Circle, about whom more later.

170   **in Holland, Italy, France, and Germany (in that order) from 1720 on:** A. Rupert Hall, "Making Sense of the Universe," *Nature* 327 (1987): 669.

170   **Russell McCormmach's poignant novel:** R. McCormmach, *Night Thoughts of a Classical Physicist* (Cambridge, Mass.: Harvard University Press, 1982).

174   **Andre Linde and other cosmologists:** This work builds on the so-called inflationary cosmology of Alan Guth.

175   **in a letter to him a few years later:** Quoted by J. Bernstein, "Ernst Mach and the Quarks," *American Scholar* 53 (winter 1983–84): 12.

175   **Heisenberg's great first paper:** This translation is taken from *Sources of Quantum Mechanics,* ed. B. L. van der Waerden (New York: Dover, 1967).

176   **George Gale even blames positivism:** G. Gale, "Science and the Philosophers."

177   **Mach wrote in a running debate:** E. Mach, *Physikalische Zeitschrift* 11 (1910): 603; trans. J. Blackmore, *British Journal of the Philosophy of Science* 40 (1989): 524. There is a debate among historians of science, reviewed by Blackmore, about whether Mach was ever philosophically reconciled to Einstein's special theory of relativity, which had been influenced by Mach's own doctrines.

178   **But Kaufmann was a positivist:** My friend Sambursky (whom I quoted in chap. 5) had as a very young man known Kaufmann. He confirmed my impression of Kaufmann as a rigid person confined by his own philosophy.

179   **observation can never be freed of theory:** This point has been forcefully made by the philosopher Dudley Shapere, "The Concept of Observation in Science and Philosophy," *Philosophy of Science* 49 (1982): 485–525.

180   **In a lecture in 1974 Heisenberg recalled:** W. Heisenberg, in *Encounters with Einstein, and Other Essays on People, Places and Particles* (Princeton, N.J.: Princeton University Press, 1983), p. 114.

180   **Einstein referred to Mach:** J. Bernstein, "Ernst Mach."

182   **In the end this program failed:** Nevertheless I think that we learned some valuable lessons from S-matrix theory. Quantum field theory is the way it is because that is the only way to guarantee that the observables of the theory and in particular the S-matrix would have sensible physical properties. In 1981 I gave a talk at the Radiation

Laboratory at Berkeley, and because I knew that Geoffrey Chew would be in the audience, I went out of my way to say nice things about the positive influence of S-matrix theory. After the talk Geoff came up to me and said that he appreciated my remarks but he was now working on quantum field theory.

183 **certain kinds of quantum field theory:** I am referring here to the so-called non-Abelian or Yang-Mills gauge theories.

183 **the forces in these theories decrease at high energies:** This calculation used mathematical methods introduced in 1954 in the context of quantum electrodynamics by Murray Gell-Mann and Francis Low. But the force in quantum electrodynamics and in most other theories increases with increasing energy.

183 **experiments on high-energy scattering going back to 1967:** Particularly experiments on the disruption of neutrons and protons by high-energy electrons carried out at the Stanford Linear Accelerator Center by a group led by Jerome Friedman, Henry Kendall, and Richard Taylor.

183 **a few theorists proposed instead:** Gross and Wilczek, and myself.

183 **It is now believed that if you try:** As far as I know, this picture is due independently to G. 't Hooft and L. Susskind. An early suggestion of quark trapping was also made by H. Fritzsch, M. Gell-Mann, and H. Leutwyler.

184 **part of the accepted wisdom of modern elementary particle physics:** The case for the existence of quarks became compelling with the discovery in 1974 by groups headed by Burton Richter and Sam Ting of a particle they respectively called the *psi* and *J* particle. The properties of this particle showed clearly that it consisted of a heavy new quark and its antiquark, even though these quarks could not be produced in isolation. (The existence of this type of heavy quark had been proposed earlier by Sheldon Glashow, John Iliopoulos, and Luciano Maiani as a means of avoiding certain problems in the theory of weak interactions, and its mass had estimated theoretically by Mary Gaillard and Ben Lee. The J-psi particle had been predicted by Thomas Appelquist and David Politzer.)

184 **The philosophical relativists deny the claim of science:** For an etiology and criticism of the relativists, see M. Bunge, "A Critical

Examination of the New Sociology of Science," *Philosophy of the Social Sciences* 21 (1991): 524 [Part 1] and ibid., 22 (1991):46 [Part 2].

184   In his celebrated book: T. Kuhn, *The Structure of Scientific Revolutions*, 2nd ed., enlarged (Chicago: University of Chicago Press, 1970).

185   many of her observations: S. Traweek, *Beamtimes and Lifetimes: The World of High Energy Physicists* (Cambridge, Mass.: Harvard University Press, 1988).

186   A recent book on peer review: D. E. Chubin and E. J. Hackett, *Peerless Science: Peer Review and U.S. Science Policy* (Albany, N.Y.: State University of New York Press, 1990); quoted in a book review by Sam Treiman, *Physics Today,* October 1991, p. 115.

186   Close observation of scientists at work at the Salk Institute: B. Latour and S. Woolgar, *Laboratory Life: The Social Construction of Scientific Facts* (Beverly Hills, Calif., and London: Sage Publications, 1979), p. 237.

186   a book by Andrew Pickering: A. Pickering, *Constructing Quarks: A Sociological History of Particle Physics* (Chicago: University of Chicago Press, 1984).

188   terms suggestive of a mere change of fashion: Similar views were expressed in the earlier writings (more than twenty years ago) by Feyerabend, but he has since changed his mind. Traweek carefully avoids this issue; she expresses sympathy with the physicists' view that electrons exist, acknowledging that in her work she finds it appropriate to assume that physicists exist.

189   a wider, radical, attack on science itself: For a collection of articles on the critics of science, see *Science and Its Public: The Changing Relationship,* ed. G. Holton and W. Blanpied (Boston: Reidel, 1976). A more recent commentary is given by G. Holton, "How to Think About the "Anti-science Phenomenon,'" *Public Understanding of Science* 1 (1992): 103.

189   Feyerabend called for a formal separation of science and society: P. Feyerabend, "Explanation, Reduction, and Empiricism."

189   "not only sexist but also": S. Harding, *The Science Question in Feminism* (Ithaca, N.Y.: Cornell University Press, 1986), p. 9.

189   "Physics and chemistry, mathematics and logic": Ibid., p. 250.

189   "the fundamental sensibility of scientific thought": T. Roszak, *Where the Wasteland Ends* (Garden City, N.Y.: Doubleday, Anchor Books, 1973), p. 375.

190   **I do not know of any working scientist who takes them seriously:** This is acknowledged by Evelyn Fox Keller, in *Reflections on Gender and Science* (New Haven: Yale University Press, 1985). (As an example of the attitude of scientists, Keller quotes an old remark of mine: "The laws of nature are as impersonal and as free of human values as the rules of arithmetic. We didn't want it to come out this way, but it did.") More recently, responding to the heavy-handed sociological reinterpretation of scientific progress, the University of London geneticist J. S. Jones remarked that "the sociology of science bears the same relation to research as pornography does to sex: it is cheaper, easier and—as it is constrained only by the imagination—can be a lot more fun" [in a review of *The Mendelian Revolution: The Emergence of Hereditarian Concepts in Modern Science and Society,* by Peter J. Bowler, *Nature* 342 (1989): 352].

190   **Recently the minister in charge of government spending on civil science in Britain:** Editorial in *Nature* 356 (1922): 729. The minister in question is George Walden, M.P.

190   **of a book by Bryan Appleyard:** B. Appleyard, *Understanding the Present* (London: Picador, 1992).

190   **I suspect that Gerald Holton is close to the truth:** G. Holton, "How to Think About the End of Science," in *The End of Science,* ed. R. Q. Elvee (Lanham, Minn.: University Press of America, 1992).

CHAPTER VIII.   TWENTIETH CENTURY BLUES

193   **electrons and W and Z particles have masses, but neutrinos and photons do not:** It is possible that neutrinos and even photons do have masses so small that they have escaped detection so far, but these masses would be very different from the masses of electrons and W and Z particles, which is not what would be expected if the symmetry among these particles were manifest in nature.

194   **One might suppose that the symmetry between the two quark types would dictate that the two masses should turn out to be equal, but this is not the only possibility:** For instance, an equation that says that the ratio of the up to the down quark masses plus the

ratio of the down to the up quark masses equals 2.5 is evidently symmetric between the two quarks. It has two solutions: in one solution the up quark mass is twice the down quark mass, and in the other the down quark mass is twice the up quark mass. It has *no* solution in which the masses are equal, because then both ratios would equal 1 and their sum would equal 2, not 2.5.

195   it spontaneously develops a magnetic field pointing in some specific direction: The direction of this magnetic field is determined by any stray magnetic field that may be present, such as the field of the earth; the important thing is that the strength of the magnetism developed in the iron is the same however weak the stray field may be. In the absence of any strong external magnetic field the direction of the magnetism is different in various different "domains" within the iron, and the magnetic fields that appear spontaneously within the individual domains cancel for the magnet as a whole. The domains can be made to line up by exposing the cooling iron to a strong external magnetic field, and the magnetization will persist even when the external magnetic field is removed.

196   a symmetry that happens to be broken in *our* universe. It is the symmetry relating the weak and electromagnetic forces: This symmetry is not entirely broken; there is a remaining unbroken symmetry (known as electromagnetic gauge invariance) that dictates that the photon must have zero mass. This remaining symmetry is itself broken in a superconductor. Indeed, that is what a superconductor is—it is in essence nothing but a piece of matter in which electromagnetic gauge invariance is broken.

197   The elusive neutrino was eventually discovered: By C. L. Cowan and F. Reines.

197   the mathematics of simpler examples of this sort of symmetry breaking had been described by a number of theorists: Including F. Englert and R. Brout, and G. S. Guralnik, C. R. Hagen, and T. W. B. Kibble.

199   it is possible that the breakdown of the electroweak symmetry is due to the indirect effects of some new kind of extra-strong force: This new force could cause *products* of the fields of any particles that feel the force to develop vacuum values, which could break the electroweak symmetry, even though the vacuum values of the individual fields are all zero. (It is a familiar feature of probabilities that a product of quantities can have a nonzero average value even when the average values of the individual quantities vanish. E.g., the average height of ocean waves above mean sea level is by definition zero, but the *square* of the height of ocean waves—i.e., the product of the height with itself—has a nonzero average value.) This new force could have escaped detection if it acts only on hypothetical particles that are too heavy to have been discovered yet.

199   Such theories were developed in the late 1970s: These theories were developed independently by Lenny Susskind of Stanford and myself. In order to distinguish the new kind of extra-strong force needed in such theories from the familiar strong "color" forces that bind quarks inside the proton, the new force has come to be called *technicolor,* a name owing to Susskind. The trouble with the technicolor idea is that it does not account for the masses of quarks, electrons, etc. It is possible to give these particles masses and to avoid conflict with experiment by various elaborations of the theory, but the theory then becomes so baroque and artificial that it is difficult to take seriously.

200   the strong as well as the weak and electromagnetic interactions would be unified: Theories that unify the strong with the electroweak interactions are often called grand unified theories. Specific theories of this sort were proposed by Jogesh Pati and Abdus Salam; Howard Georgi and Sheldon Glashow; and H. Georgi; and since then by many others.

201   In 1974 an idea appeared: This was the work of Howard Georgi, Helen Quinn, and myself.

201   one prediction that related the strengths: More precisely, it is just one ratio of these strengths that is predicted. When this prediction was made in 1974 it seemed at first like a failure; this ratio was predicted to be 0.22, but experiments on neutrino scattering showed that instead it had a value of about 0.35. As time has passed since the mid-1970s the experimental value for this ratio has decreased, and it is now quite close to the expected value of 0.22. But both measurements and theoretical calculations are now so accurate that we can see that there is a discrepancy of several percent between them. As we shall see, there are theories (embodying the symmetry known as supersymmetry) that resolve this remaining discrepancy in a very natural way.

205   a new sort of symmetry, known as *supersymmetry*: Supersymmetry was introduced as a fascinating possibility by Julius Wess and Bruno Zumino in 1974, but its potential for solving the hierarchy problem has been responsible for much of the interest in supersymmetry since then. (Versions of supersymmetry had already appeared in earlier papers by Yu. A. Gol'fand and E. P. Likhtman and by D. V. Volkov and V. P. Akulov, but its physical significance had not been explored in these papers, and it attracted little attention. Wess and Zumino drew at least part of their inspiration from work on string theory in 1971 by P. Ramond, A. Neveu and J. H. Schwarz, and J.-L. Gervais and B. Sakita.)

205   the symmetry forbids the appearance of any Higgs particle masses in the fundamental equations of the theory: Until the advent of supersymmetry it was thought that it would be impossible for any symmetry to forbid such masses. The absence of masses for particles like quarks and electrons and the photon, W and Z particles, and gluons in the equations of the original version of the standard model is inseparably connected with the fact that these particles have spin. (The familiar phenomenon of polarized light is a direct effect of the spin of the photon.) But in order for a field to have a nonzero vacuum value that breaks the electroweak symmetry, the field must not have any spin; otherwise its vacuum value would also break the symmetry of the vacuum with regard to changes of direction, in gross contradiction to experience. Supersymmetry solves this problem by establishing a relation between a spinless field whose vacuum value breaks the electroweak symmetry and the various fields that have spin and are forbidden by the electroweak symmetry from having any masses in the field equations. Supersymmetry theories have their own problems: The superpartners of the known particles have not been discovered, so they must be much heavier, and thus supersymmetry itself must be a broken symmetry. There are various interesting suggestions for the mechanism that breaks supersymmetry, some of them involving the force of gravity, but so far the question is open.

205   In another approach . . . the effects of some new extra-strong force: A version of the standard model that is based on the introduction of new extra-strong (technicolor) forces would avoid the hierarchy problem because there would be no masses at all in the equations that describe physics at energies far below the Planck energy. The scale of masses of the W and Z particles and the other elementary particles of the standard model would be related instead to the way that the strength of the technicolor force changes with energy. The technicolor force as well as the strong and electroweak forces would be expected to have the same intrinsic strength at some very high energy, not very different from the Planck energy. With decreasing energy its strength would increase very slowly, so that the technicolor force would not become strong enough to break any symmetries until the energy drops to a value very much lower than the Planck energy. It is quite plausible that without any fine-tuning of the constants of the theory the technicolor force would strengthen with decreasing energy a little faster than the ordinary color force, so that it could give something like the observed masses for the W and Z particles of the standard model, while the ordinary color force acting alone would give them masses a thousand times smaller.

206   Unfortunately there is so far no sign of supersymmetry: Supersymmetry calls for all the known quarks and photons and so on to have "superpartners" of different spin. Even though none of these have been seen, theorists have not been slow to give names to all these particles: the superpartners (with zero spin) of particles like the quarks, electrons, and neutrinos, are called squarks, selectrons, sneutrinos, and so on, while the superpartners (with half the spin) of the photon, W, Z, and gluons are called the photino, wino, zino, and gluinos. I once proposed to call this patois a "languino," but Murray Gell-Mann has suggested a better term: it is a "slanguage." Just recently the idea of supersymmetry has received an important boost from experiments on the decay of the Z particle at the CERN Laboratory in Geneva. As mentioned earlier, these experiments are now so accurate that it is possible to tell that there is a small discrepancy (about 5%) between the interaction strength ratio of 0.22 predicted in 1974 and the actual value. Interestingly, calculations show that the presence of squarks and gluinos and all the other new particles required by supersymmetry would change the way that the interaction strengths change with energy just enough to bring theory and experiment back into agreement.

209   fewer neutrinos than expected are detected to be coming from the sun: This was first observed in 1968, in the comparison of experimental results of Ray Davis, Jr., with the calculation of the expected flux of neutrinos by John Bahcall.

209   perhaps electron-type neutrinos seem to be missing because as they pass through the sun they have turned into neutrinos of other types: This was suggested in 1985 by S. P. Mikhaev and A. Yu. Smirnov, on the basis of earlier work by Lincoln Wolfenstein.

CHAPTER IX.   THE SHAPE OF A FINAL THEORY

213   In the course of this work it was realized: Independently by Yoichiro Nambu, Holger Nielsen, and Leonard Susskind.

214   there is no place for their energy of vibration to go: This remark is due to Edward Witten.

214   The early versions of string theory were not without problems: Some of these difficulties could be avoided only by imposition of the symmetry that was later called supersymmetry, so that these are often called *superstring* theories.

214   there was one mode in which the string would appear like a particle with zero mass and a spin twice that of the photon: Although this unwanted particle appeared in string theories as a mode of vibration of a *closed* string, it would not have been possible to avoid the appearance of this particle by considering only open strings because colliding open strings inevitably join up to form closed strings.

215   any theory of a particle with this spin and mass would have to look more or less the same as general relativity: This conclusion was reached independently by Richard Feynman and myself.

215   the new massless particle . . . was in fact the true graviton: This had first been suggested as early as 1974 by J. Scherk and J. Schwarz and independently by T. Yoneya.

216   "the greatest intellectual thrill of my life": Quoted by John Horgan in *Scientific American*, November 1991, p. 48.

216   string theories seem to be free of any infinities: It is true that a string theory can be regarded as just a theory of particles corresponding to the different modes of vibration of the string, but because of the infinite number of species of particles in any string theory, string theories work differently from ordinary quantum field theories. E.g., in a quantum field theory the emission and reabsorption of a single species of particle (such as a photon) produces an infinite energy shift; in a properly formulated string theory this infinity is canceled by effects of the emission and absorption of particles belonging to the infinite number of other species present in the theory.

217   a test for mathematical consistency that had been failed: This inconsistency in some string theories had been discovered a little earlier by Witten and Luis Alvarez-Gaumé.

217 **one team of theorists:** Philip Candelas, Gary Horowitz, Andrew Strominger, and Edward Witten.

217 **the "Princeton String Quartet":** David Gross, Jeffrey Harvey, Emil Martinec, and Ryan Rohm.

217 **conformal symmetry seems to be necessary:** Conformal symmetry is based on the fact that a set of strings as they move through space sweep out a two-dimensional surface in space-time: each point on the surface has one coordinate label giving the time and another label specifying location along one of the strings. Just as for any other surface, the geometry of this two-dimensional surface swept out by the strings is described by specifying the distances between any pair of very close points in terms of their coordinate labels. The principle of conformal invariance states that the equations governing the string preserve their form if we change the way we measure distances by multiplying all the distances between one point and any adjacent point by an amount that may depend in an arbitrary way on the position of the first point. Conformal symmetry is required because the vibrations of the string in the time direction would otherwise lead (according to one's formulation of the theory) either to negative probabilities or to vacuum instability. With conformal symmetry these timelike vibrations can be removed from the theory by a symmetry transformation, and are thus innocuous.

220 **the *anthropic principle:*** The term "anthropic principle" is due to Brandon Carter; see *Confrontation of Cosmological Theories with Observation,* ed. M. S. Longair (Dordrecht: Reidel, 1974). Also see B. Carter, "The Anthropic Principle and Its Implications for Biological Evolution," in *The Constants of Physics,* ed. W. McCrea and M. J. Rees (London: Royal Society, 1983), p. 137; reprinted in *Philosophical Transactions of the Royal Society of London* A310 (1983): 347. For a thorough discussion of various versions of the anthropic principle, see J. D. Barrow and F. J. Tipler, *The Anthropic Cosmological Principle* (Oxford: Clarendon Press, 1986); J. Gribbin and M. Rees, *Cosmic Coincidences: Dark Matter, Mankind, and Anthropic Cosmology* (New York: Bantam Books, 1989), chap. 10; J. Leslie, *Universes* (London: Routledge, 1989).

220 **The solution found eventually by Edwin Salpeter:** Salpeter in his 1952 article also credits E. J. Öpik with having the same idea in 1951.

220   **experimenters working with Hoyle:** D. N. F. Dunbar, W. A. Wensel, and W. Whaling.

220   **no obstacle to building up all the heavier elements:** In fact, the energy levels of oxygen must also have certain special properties in order to avoid all of the carbon being cooked into oxygen.

221   **a group of physicists:** M. Livio, D. Hollowell, A. Weiss, and J. W. Truran.

221   **the energy of the state of carbon in question could be increased appreciably:** Specifically, by about 60,000 volts. This is admittedly a very small energy compared with the difference of 7,644,000 volts between the energy of this unstable state and that of the stable lowest energy state of carbon. But it does not take any fine-tuning to make the energy of this unstable state of the carbon nucleus equal within this accuracy to the energy of a nucleus of beryllium 8 and a helium nucleus, because to a good approximation the relevant states of the carbon and beryllium nuclei are just loosely bound nuclear molecules consisting of three or two helium nuclei. (I thank my colleague Vadim Kaplunovsky of the University of Texas for this remark.)

221   **there is a context in which it would be only common sense:** This version of the anthropic principle is sometimes known as the weak anthropic principle.

222   **One very simple possibility proposed by Hoyle:** F. Hoyle, *Galaxies, Nuclei, and Quasars* (London: Heinemann, 1965).

222   **"wormholes" can open up:** Strictly speaking, these wormholes appear mathematically in an approach to quantum gravity known as Euclidean path integration. It is not clear what they have to do with actual physical processes.

222   **in each of which the "constants" of nature take different values, with various different probabilities:** Coleman went on to argue also (as Baum and Hawking had earlier) that the probabilities for these constants have infinitely sharp peaks at certain special values, so that it is overwhelmingly likely that the constants would take these special values. But this conclusion is based on a mathematical formulation (Euclidean path integration) of quantum cosmology whose consistency has been called into question. It is difficult to be certain about such matters because we are dealing with gravitation in a quantum context where our present theories are no longer adequate.

224 **Einstein came to regret mutilating his equations:** To show once again how complicated the history of science can be, I will mention that immediately after Einstein's 1917 work on cosmology his friend Wilhelm de Sitter pointed out that Einstein's gravitational field equations when modified by the inclusion of a cosmological constant have a different class of solutions, also apparently static, but containing no matter (or negligible matter). This was disappointing to Einstein because in his solution the cosmological constant is related to the average cosmic density of matter, in accordance with what Einstein took to be Mach's teachings. Furthermore the Einstein solution (with matter) is actually unstable; any small perturbation would cause it eventually to turn into the de Sitter solution. To complicate matters even more, I will point out that the de Sitter model is only apparently static; although the space-time geometry in the coordinate system used by de Sitter does not change with time, any small test particles put into his universe rush apart from one another. In fact, when Slipher's measurements became known in England in the early 1920s, they were at first interpreted by Arthur Eddington in terms of the de Sitter solution of the Einstein equations *with* a cosmological constant, which also has a static solution, rather than in terms of the original Einstein theory, which does not!

225 **Theoretical physicists have been trying for years to understand the cancellation of the total cosmological constant:** A nonmathematical account is given by L. Abbott, *Scientific American* 258, no. 5 (1985): 106.

226 **in the general case this comes out enormously too large:** We cannot even hope that some mechanism will be found by which the vacuum state can lose its energy by decaying into a state of lower energy and hence lower total cosmological constant and ultimately winding up in a state with zero total cosmological constant, because some of these possible vacuum states in string theories already have a large *negative* total cosmological constant.

227 **The most natural value for the mass density of the universe:** The discovery of any lower or higher density would raise the question why the expansion has continued for billions of years and yet is still slowing down.

CHAPTER X. FACING FINALITY

230   rejects "the idea of an ultimate explanation": K. R. Popper, *Objective Knowledge: An Evolutionary Approach* (Oxford: Clarendon Press, 1972), p. 195.

232   The Cambridge philosopher Michael Redhead suggests: M. Redhead, "Explanation," August 1989, to be published.

233   the suggestion that at bottom we shall find that there is no law at all: An interesting discussion of this possibility has been given by Paul Davies, "What Are the Laws of Nature," in *The Reality Club #2*, ed. John Brockman (New York: Lynx Communications, 1988).

233   John Wheeler has occasionally suggested: See, e.g., J. A. Wheeler, "On Recognizing 'Law Without Law,'" Oersted Lecture presented at the Joint Ceremonial Session of the American Association of Physics Teachers and the American Physical Society, January 25, 1983, *American Journal of Physics* 51 (1983): 398. J. A. Wheeler, "Beyond the Black Hole," in *Some Strangeness in the Proportion: A Centennial Symposium to Celebrate the Achievements of Albert Einstein*, ed. H. Woolf (Reading, Mass.: Addison-Wesley, 1980), p. 341.

233   Holger Nielsen has proposed a "random dynamics": H. B. Nielsen, "Field Theories Without Fundamental Gauge Symmetries," in *The Constants of Physics*, ed. W. McCrea and M. J. Rees (London: Royal Society, 1983), p. 51; reprinted in *Philosophical Transactions of the Royal Society of London* A310 (1983): 261.

234   Eugene Wigner has warned: E. P. Wigner, "The Limits of Science," *Proceedings of the American Philosophical Society* 94 (1950): 422.

236   Redhead probably represents a majority view: M. Redhead, "Explanation."

236   The Harvard philosopher Robert Nozick: R. Nozick, *Philosophical Explanation* (Cambridge, Mass.: Harvard University Press, 1981), chap. 2.

CHAPTER XI. WHAT ABOUT GOD?

241   "The heavens declare": Psalms 19:1 (King James version).

242   So it is natural that Stephen Hawking should refer to the laws of nature as "the mind of God.": S. Hawking, *A Brief History of*

Time (London: Bantam Books, 1988); I have also seen titles of two recent books that use the same expression: J. Trefil, *Reading the Mind of God* (New York: Scribner, 1989), and P. Davies, *The Mind of God: The Scientific Basis for a Rational World* (New York: Simon & Schuster, 1992).

242   **Another physicist, Charles Misner, used similar language:** C. W. Misner, in *Cosmology, History, and Theology,* ed. W. Yourgrau and A. D. Breck (New York: Plenum Press, 1977), p. 97.

242   **Einstein once remarked to his assistant:** A. Einstein, quoted by Gerald Holton in *The Advancement of Science, and Its Burdens* (Cambridge: Cambridge University Press, 1986), p. 91.

242   **On another occasion, he described the aim of the enterprise of physics:** A. Einstein, contribution to *Festschrift für Aunel Stadola* (Zurich: Orell Füssli Verlag, 1929), p. 126.

242   **The theologian Paul Tillich once observed:** P. Tillich, in a talk at the University of North Carolina, c. 1960, quoted by B. De Witt, "Decoherence Without Complexity and Without an Arrow of Time," University of Texas Center of Relativity preprint, 1992.

243   **There then ensued a dialogue between two committee members:** This is from the unedited transcript of the hearings. Congressmen, unlike witnesses, have the privilege of editing their remarks for the *Congressional Record.*

245   **Einstein once said that he believed in:** Interview in the *New York Times,* April 25, 1929. I am grateful to A. Pais for this quotation.

245   **Galileo, who made it plausible that Copernicus was right:** Galileo's work on motion showed that we on earth would not feel the earth's motion around the sun. Also, his discovery of moons circling Jupiter provided an example of a sort of solar system in miniature. The crowning proof came with the discovery of the phases of Venus, which did not match what would be expected if both Venus and the sun revolve about the earth.

245   **the same law of gravitation governs the motion of the moon around the earth and a falling body:** The moon in circling about the earth instead of flying off in a straight line to outer space in effect acquires a component of velocity of a tenth of an inch per second toward the earth in each second. Newton's theory explained that this is thirty-six hundred times less than the acceleration of a falling apple in Cam-

bridge because the moon is sixty times farther than Cambridge is from the center of the earth and the acceleration due to gravity decreases as the inverse square of the distance.

247   **Schrödinger's mistake was pointed out by Max Perutz:** M. F. Perutz, "Erwin Schrödinger's *What Is Life?* and Molecular Biology," in *Schrödinger: Centenary Celebration of a Polymath,* ed. C. W. Kilmeister (Cambridge: Cambridge University Press, 1987), p. 234.

247   **Professor Phillip Johnson:** I first learned of Professor Johnson when a friend passed on to me his article, "Evolution as Dogma," in *First Things: A Monthly Journal of Religion and Public Life,* October 1990, pp. 15–22. He has also recently published a book, *Darwin on Trial* (Washington, D.C.: Regnery Gateway, 1991), and according to a story in *Science* 253 (1991): 379, he is busy on the lecture circuit publicizing his views and writings.

249   **in a review of Johnson's book:** S. Gould, "Impeaching a Self-Appointed Judge," *Scientific American,* July 1992, p. 118.

254   **John Polkinghorne has argued eloquently for a theology:** J. Polkinghorne, *Reason and Reality: The Relation Between Science and Theology* (Philadelphia: Trinity Press International, 1991).

255   **that phrase has dogged me ever since:** For two recent comments, see S. Levinson, "Religious Language and the Public Square," *Harvard Law Review* 105(1992): 2061; M. Midgley, *Science as Salvation: A Modern Myth and Its Meaning* (London: Routledge, 1992).

255   **interviews with twenty-seven cosmologists and physicists:** A. Lightman and R. Brawer, *Origins: The Lives and Worlds of Modern Cosmologists* (Cambridge, Mass.: Harvard University Press, 1990).

257   **To borrow a phrase from Susan Sontag:** S. Sontag, "Piety Without Content," in *Against Interpretation and Other Essays* (New York: Dell, 1961).

258   **The historian Hugh Trevor-Roper has said:** H. R. Trevor-Roper, *The European Witch-Craze of the Sixteenth and Seventeenth Centuries, and Other Essays* (New York: Harper & Row, 1969).

259   **As Karl Popper has said:** K. R. Popper, *The Open Society and Its Enemies* (Princeton, N.J.: Princeton University Press, 1966), p. 244.

259   **David Hume saw long ago:** See his *Treatise on Human Nature* (1739).

261   *The Ecclesiastical History of the English:* Bede, *A History of the English Church and People,* trans. Leo Sherley-Price and rev. R. E. Latham (New York: Dorset Press, 1985), p. 127.

CHAPTER XII.   DOWN IN ELLIS COUNTY

265   **Afterward Samios called this vote:** Quoted in *Science* 221 (1983): 1040.

266   **announced his department's decision to stop work on ISA-BELLE:** The ISABELLE tunnel is now to be used for the Relativistic Heavy Ion Collider, an accelerator that will be used to study collisions of heavy atomic nuclei, with the aim of understanding nuclear matter rather than the fundamental principles of elementary particle physics. This heavy ion collider is expected to be ready in 1997.

268   **exotic dark matter . . . would therefore have started its grav-itational condensation much closer to the beginning:** This remark applies to nonuniformities of galactic size but not to the much larger nonuniformities inferred from the COBE measurements. These are so large that even a light wave could not have crossed them during the first three hundred thousand years after the beginning of the present expansion of the universe, and therefore (whether or not composed of dark matter) they could not have experienced any significant growth in this time.

271   **every year the same arguments are made for and against it:** After the Ellis County site had been chosen, a new element entered the debate: the charge by disappointed politicians from states like Arizona, Colorado, and Illinois that Texas had won the site competition by unfair political pressures. It was widely remarked that the Department of Energy's choice of a Texas site for the SSC was announced just two days after the election of George Bush of Texas to the presidency. Secretary of Energy Herrington said after the SSC site decision was announced that the Department of Energy task force that rated the seven "highly qualified" sites was insulated from political pressures; that he himself did not receive their briefing until election day; that the task force rated the Texas site as clearly superior; and that only then did he clear the final decision with President Reagan and President-elect Bush. I can well believe that the process could have been speeded up and the decision announced before the election, but then it doubtless would have been charged that the announcement was timed to influence the important Texas vote. On the other hand, even if the site selection had not been affected by the election of George Bush, the Department of Energy certainly had known all along of the strength of the Texas congressional delegation and its enthusiasm for the SSC, and may have hoped that a decision for a Texas site would improve the SCC's chances for funding by Congress. If so, it would hardly be a scandal, or the first or the last time that such calculations have been made by a government agency. In any case, I can testify that calculations of this sort played no part in the selection of the seven highly qualified sites by the National Academies committee on which I served. Our committee had from the beginning regarded the Texas site as one of the leading contenders. This was in part because of its exceptionally good geology. Another important factor was the vocal local opposition to the SSC at several of the other best-qualified sites, including the one at Fermilab in Illinois. In Ellis County almost everyone was glad to welcome the SSC.

271   a **"quark-barrel" project:** D. Ritter, *Perspectives,* summer 1988, p. 33

273   the **freed funds have been allocated to water projects:** See, e.g., R. Darman, quoted by P. Aldhous in "Space Station Back on Track," *Nature* 351 (1991): 507.

INDEX